Second Edition

Programming of Computer Numerically Controlled Machines

Second Edition by

Ken Evans

John Polywka *and* Stanley Gabrel

Library of Congress Cataloging-in-Publication Data

Evans, Ken (Kenneth W.)
   Programming of computer numerically controlled machines / second edition by Ken Evans; John Polywka and Stanley Gabrel.--2nd ed.
   p.cm.
   Previous edition under John Polywka and Stanley Gabrel.
   Includes index.
   ISBN 0-8311-3129-2
   1. Machine-tools—Numerical control—Programming. I. Polywka, John. II. Gabrel, Stanley. III. Title.
TJ1189.E85 2001
621.9'023'028551—dc21

2001016988

INDUSTRIAL PRESS INC.
200 Madison Avenue, New York, NY 10016

Programming of Computer Numerically Controlled Machines
Second Edition

Copyright © 2001 by Industrial Press Inc., New York, New York. Printed in the United States of America. All rights reserved. This book, or parts thereof, may not be reproduced, stored in a retrieval system, or transmitted in any form without the permission of the publishers.

Notice to the reader: While every possible effort has been made to insure the accuracy of the information presented herein, the authors and publisher express no guarantee of the same. The authors and publisher do not offer any warrant or guarantee that omissions or errors have not occurred and may not be held liable for any damages resulting from the use of this text by the readers. The readers accept full responsibility for their own safety and that of the equipment used in connection with the instructions in this text. There has been no attempt to cover all controllers or machine types used in the industry and the reader should consult the operation and programming manuals of the machines they are using before any operation or programming is attempted.

10 9 8 7 6 5 4 3 2

Dedication

To my loving wife, Marci, for her support, patience, and proofreading skills. She kept me on track and offered countless valuable critiques of my machinist's English. Thanks also to my entire family for allowing the use of our home for this effort.

Acknowledgements

I give thanks first to the Lord, Our God, for blessing me with the opportunity, knowledge and ability to share in this work. Many thanks are due to all of the parties who helped on this project. Special thanks are due to the publisher, Industrial Press, specifically to President Alex Luchars, Marketing Director/Editor, John Carleo, Production Manager, Janet Romano and former Assistant Editor Sheryl Levart. Thanks also to Marc Sullivan of CNC Software, Inc, for his graciousness in allowing the use of Mastercam software in the development of this work. Richard Jones of Technical Training Systems, Lab Technologies for his graciousness in allowing the use of AutoCad 14 software in the development of this work. Peter Smid for his thoughtful efforts in the review of the work.

Thanks to:

Kevin Cummings, Planning and Development Coordinator, of the Davis Applied Technology Center in Kaysville, Utah for his technical advice.

Sandvik Coromant, for their contribution of tooling drawings and other technical data.

CarrLane Manufacturing, for their permission to use technical data in the appendix.

Mazak Corporation, for their contribution of photos used in the Conversational Programming section.

CNC Software Inc., Mastercam, for their permission to use screen shots of the programming process in the CAD/CAM section.

GE Fanuc, for the use of drawings of their controller, operation panel and screen shots of the set-up, operation and programming displays.

The original authors John Polywka and Stanley Gabrel for their efforts in building a solid foundation in the first edition of this book.

Foreword

It has been said that learning is a lifetime process. In the rapidly evolving computer age, this has never been more true. Manufacturing in general, and machining in particular, has not been immune from the growth of new technologies. CNC programming and CNC machining have not remained untouched, as new materials, new tools, new machine and control features are introduced to the industry. Good learning material to unravel the new approaches and techniques is hard to find. This second edition of "Programming of Computer Numerically Controlled Machines" has successfully attempted to fill many voids. As a complete rewrite by Ken Evans of the popular book of the same name by Stanley Gabrel and John Polywka, the book approaches the subject of CNC with 21st century manufacturing in mind. This is a book that has it all.

The best features of the book are its contents and style. The book is very easy to understand – the author shows his skill as a professional communicator on every page. His extensive experience in both industry and educational fields give him a high level of credibility. He tries to be original and, without a doubt, succeeds very well.

The book presents the subject of CNC programming in a practical and well-organized way. Numerous examples, study questions, charts and mathematical formulae complement the extensive text. Illustrations throughout the book lead the reader to the subject of interest.

Written for machinists with little or no CNC experience, this book is a valuable resource for learning CNC programming. The "Operation" section in the early part of the book is designed to ease an experienced machinist into the world of CNC programming. Programming examples are practical, well documented and selected as being typical in machine shops. At the end of the book, the Glossary, the Appendix, and the Index can be easily accessed for instant reference.

As a major update of a popular book, the second edition of "Programming of Computer Numerically Controlled Machines" will undoubtedly find its way as a CNC resource for the thousands of machinists, programmers and managers.

Peter Smid
Author of "CNC Programming Handbook"

Preface

In this second edition, you will notice many changes and improvements. The Operation section has been moved to the second chapter, because most machinists are first exposed to CNC as an operator. They will not be concerned right away with programming. After more on-the-job time and practice, and with the confidence of the owner, operators are given greater responsibilities such as changing wear offsets, performing set-ups and minor editing of the program. In this first and second chapters of the new edition, emphasis regarding program editing will reflect this position. Engineers will benefit by learning about the operation of a machine tool prior to programming it. Because of the use of Computer Aided Design and Computer Aided Manufacturing (CAD/CAM) and Conversational Programming at the machine controller, two new chapters have been added. Because of the reduced cost and the effectiveness of CAD/CAM, it is now the conventional method for programming today. Many new machine tools come standard with some form of Conversational Programming. All of the drawings have been recreated to ANSI standards in order to improve clarity. The appendix contains many useful charts and math formulas used for programming and a glossary of terms has been added.

The purpose of this book is to expand the current knowledge of CNC programming by providing full descriptions of all program functions and their practical applications. The book contains information on how to program turning and milling machines, which is applicable to almost all control systems. In order to provide clear explanations about one unified system, we chose as our model one of the most widely accepted and popular numerical control systems used worldwide.

The authors of this book have been active programmers with a strong interest in the practical application of Computer Numerical Control. Therefore, all theoretical explanations are kept to a minimum so that they do not distort an understanding of the programming. Because of the wide range of information available about the selection of tools, cutting speeds, and the technology of machining, we want this book to reach a wide range of readers. Included among these are persons already involved in programming or maintaining CNC machines, operators of conventional machines who may want to expand their knowledge beyond conventional machining, and managers or other interested persons who may wish to purchase such machines in the near future. Finally, we hope those with no experience in this field whatsoever will find our book to be of interest to them as well.

Ken Evans
Stanley Gabrel
John Polywka

Table of Contents

Dedication	*iii*
Acknowledgements	*iv*
Foreward	*v*
Preface	*vi*
PART 1 GENERAL INFORMATION	**3**
Safety	3
Maintenance	4
Tool Clamping Methods	5
Cutting Tool Selection	6
Tool Compensation Factors	7
Tool Changing	7
Metal Cutting Factors	8
Process Planning for CNC	11
Types of Numerically Controlled Machines	15
What is Programming?	16
Orientation of the Coordinate System	17
Coordinate Systems	19
Points of Reference	24
Program Format	29
PART 1 STUDY QUESTIONS	**31**
PART 2 OPERATION	**33**
Panel Descriptions	35
Operator Panel Features	35
Feedrate Override	36
Emergency Stop	36
Program Protect	36
Program Source	36

Operation Select	37
Execution	39
Operation	40
Speed/Multiply	41
Spindle	44
Axis/Direction	44
Coolant	45

CONTROL PANEL DESCRIPTIONS	**46**
Control Panel	46
Power-ON & Power-OFF	47
CRT Display	47
Soft Keys	47
Address and Numeric Keys (Alpha-Numerical Keys)	48
Part Program Edit Keys	49
Function Keys	50
Operations Performed At The CNC Control	51
MDI Operations	55
Measuring Work Offsets, Turning Center	57
Measuring Work Offsets Machining Center	57
Tool Offsets are Measured	59
Adjusting Wear-Offsets for Turning Centers	61
Machining Center Tool Offsets	63
Adjusting Wear-Offsets for Machining Centers	64
Tool Path Verification of the Program	65
Dry Run of Program	65
Execution in Automatic Cycle Mode	65
DNC Operation	66
Program Editing Functions	67

SETTING	**70**
PARAM	**70**
DGNOS	**71**
Tape Code	71
Common Operation Procedures	71

PART 2 STUDY QUESTIONS	**77**

PART 3 PROGRAMMING COMPUTER NUMERICALLY CONTROLLED TURNING CENTERS — 81

    Miscellaneous Functions (M-Codes) — 80-83
    Preparatory Functions (G-Codes) — 83-84

TOOL FUNCTION — 85
PRACTICAL APPLICATION OF TOOL WEAR OFFSET — 85
FEED FUNCTION — 87
SPINDLE FUNCTION — 88
    Constant Cutting Speed — 88

PROGRAMMING OF CNC LATHES IN ABSOLUTE AND INCREMENTAL SYSTEMS — 90
    Programming in Absolute Systems — 90
    Programming in Incremental Systems — 91
    Setting Absolute Zero of the Coordinate System (G50) — 92

PROGRAM CREATION — 96
    Program Number — 97
    Block Composition — 97
    Block Number — 98
    Part Program — 98
    Subprogram — 98
    Program Example — 99

PREPARATORY FUNCTIONS (G Functions) — 102
    Rapid Traverse Function (G00) — 102
    Linear Interpolation (G01) — 104
    Circular Interpolation (G02 and G03) — 104
    Dwell (G04) — 108
    Automatic Reference Point Return Check (G27) — 109
    Automatic Reference Point Return (G28) — 109
    Return From Reference Point (G29) — 110
    Thread Cutting (G32) — 112
    Thread Cutting Cycle (G92) — 114
    Tapered Thread Cutting using Cycle (G92) — 117
    Fixed Cutting Cycle A (G90) — 117
    Fixed Cutting Cycle B (G94) — 118

MULTIPLE REPETITIVE CYCLES — 119
- Rough Cutting Cycle (G71) — 119
- Face Cutting Cycle (G72) — 124
- Pattern Repeating (G73) — 125
- Finishing Cycle (G70) — 127
- Peck Drilling Cycle (G74) — 129
- Groove Cutting Cycle (G75) — 131
- Multiple Thread Cutting Cycle (G76) — 132

PRORGAMMING FOR THE TOOL NOSE RADIUS — 134
- Programming the Center of the Tool Nose Radius — 137
- Programming Using the Two Initial Points — 143

APPLICATION OF TOOL NOSE RADIUS COMPENSATION(TNRC) G41, G42 and G40 — 147
- Tool Nose Radius and Tip Orientation — 147
- Calling G41 or G42 in the Program — 148

PROGRAMMING EXAMPLES FOR LATHES — 149
- Description of Cutting Tools Used in Programming Examples — 150
- Application For Functions G00 And G01 In Both Absolute And Incremental Systems — 152
- Eliminating Taper In Turning — 155
- Eliminating Taper In Threading — 157
- Subprogram Application — 158
- Example Of Making A Taper Thread — 159
- Example Of Turning With Bar Stock as the Material — 163
- Example Of Long Shaft Turning — 164
- Programming Example For Making A Bushing — 169
- Example Illustrating The Application Of Functions G72 And G75 — 178
- Complex Program Example — 180
- Example Of Cutting A Three-Start Thread — 186
- Example Of Threading With A Common Tap — 186
- Example Illustrating The Application Of The Tool Nose RadiusCompensation (G41 and G42) — 187

PART 3 STUDY QUESTIONS — 189

PART 4 PROGRAMMING COMPUTER NUMERICALLY CONTROLLED MACHINING CENTERS   191

TOOL FUNCTION (T-Word)   193
Tool Changes   193

FEED FUNCTION (F-Word)   194
SPINDLE SPEED FUNCTION (S-Word)   195
PREPARATORY FUNCTIONS (G-Codes)   195
MISCELLANEOUS FUNCTIONS (M Functions)   198
PROGRAMMING IN ABSOLUTE AND INCREMENTAL SYSTEMS   199
Absolute Programming (G90)   199
Incremental Programming (G91)   199

PROGRAMMING OF ABSOLUTE ZERO POINT (G92)   200
Setting G92   201

PROGRAM CREATION   203
Program Number   203
Comments   203
Block Number   203

SUBPROGRAM   203
PROGRAM END   206
THE LINK BETWEEN FUNCTIONS G92 AND G43   206
EXPLANATION OF THE SAFETY BLOCK   211
OVERVIEW OF PREPARATORY FUNCTIONS (G Functions)   212
Rapid Traverse Positioning (G00)   212
Linear Interpolation (G01)   213
Circular Interpolation (G02, G03)   214
Helical Interpolation using G02 or G03   219
Dwell (G04)   222
Exact Stop (G09)   224
Polar Coordinate Cancellation (G15)   224
Polar Coordinate System (G16)   225
Plane Selection (G17, G18, G19)   225
Input In Inches (G20) Input In Millimeters (G21)   227
Stored Stroke Limit (G22, G23)   227
Reference Point Return Check (G27)   227

Return To The Reference Point (G28)	227
Return From The Reference Point (G29)	228
Return To Second, Third, And Fourth Reference Points (G30)	229
Cutter Compensation (G40, G41, G42)	229
Tool Length Compensation (G43, G44, G49)	240
Offset Amount Input By The Program (G10)	241
Work Coordinate Systems (G54, G55, G56, G57, G58, G59)	241
Single-Direction Positioning (G60)	244

CANNED CYCLE FUNCTIONS 244

High Speed Peck Drilling Cycle (G73)	246
Left Handed Tapping Cycle (G74)	247
Fine Boring Cycle (G76)	247
Canned Drilling Cycle Cancellation (G80)	248
Drilling Cycle, Spot Drilling (G81)	248
Counter Boring Cycle (G82)	250
Deep Hole Peck Drilling Cycle (G83)	252
Tapping Cycle (G84)	254

BORING CYCLES 257

G85 Reaming Cycle	257
G86 Boring Cycle	257
G87 Boring Cycle	261
G88 Boring Cycle	261
G89 Boring Cycle	262

EXAMPLES OF PROGRAMMING COMPUTER NUMERICALLY CONTROLLED MACHINING CENTERS 263

Absolute (G90) Or Incremental (G91) Programming Comparison	263
Selection Of Coordinate System (G92)	264
Complex Program Example 1	265
Example Illustrating The Application Of The Mirror Image	280
Example Of A Program With The Application Of A Rotation With Respect To The X Axis	281
Example Of Programming A Horizontal Milling Machine	284
Complex Program Example 2	292
Milling Example	300

Example Illustrating Application Of Function With Radius Compensation	**305**
Example Program For Drilling Of 1000 Holes Using Only Six Blocks Of Information	**307**
Example Illustrating Application Of Mathematical Formulas	**308**

PART 4 STUDY QUESTIONS   314

PART 5 COMPUTER AIDED DESIGN AND COMPUTER AIDED MANUFACTURING   319

What is CAD/CAM?	321
Geometry Creation	323
Job Setup	324
Tool Path	332
Contour (2D)	334
Verification	337
Post Processing	337
Associativity	338

PART 6 CONVERSATIONAL PROGRAMMING   339

What is Conversational Programing?	341
Turning Center Program Creation	342
Commom Data Process	342
Machining Process	344
Sequence Data	345
End	345
Machining Center Program Creation	346
Common Data Unit	346
Coordinate System	348
Machining Unit	348
Sequence Data	349
End Unit	351
The Future of CNC Programming	351
APPENDIX	**353**
GLOSSARY	**375**
INDEX	**383**
ANSWER KEY TO STUDY QUESTIONS	**386**

PART 1

General Information

General Information

Safety

You should not proceed to operate any machine without first understanding the basic safety procedures necessary to protect yourself and others from injury and the equipment from damage. Most CNC machines are provided with a number of safety devices (door interlocks, etc.) to protect personnel and equipment from injury or damage. However, operators should not rely solely on these safety devices, but should operate the machine only after reading and fully understanding the Safety Precautions and Basic Operating Practices outlined in the manuals provided with the equipment.

Safety Rules for NC and CNC Machines

Do's:

- Wear safety glasses and safety shoes at all times.
- Know how to stop the machine under emergency conditions.
- Keep the surrounding area well lighted, dry, and free from obstructions.
- Keep hands out of the path of moving parts, during machining operations.
- All setup work must be performed with the spindle stopped.
- Follow recommended safety policies and procedures when operating machinery, handling parts or tooling, and when lifting.
- Machine guards should be in position during operation.
- Wrenches, tools, and parts should be kept off the machine's moving parts.
- Make sure fixtures and workpieces are securely clamped before starting the machine.
- Workpieces should be loaded and unloaded with the spindle stopped.
- Cutting tools should be inspected for wear or damage prior to use.

Don'ts:

- Never operate a machine until properly instructed in its use.
- Never wear neckties, long sleeves, wristwatches, rings, gloves or long hair loose, when operating any machine.

Part 1 General Information

- Never attempt to remove chips with fingers.
- Never direct compressed air at yourself or others.
- Never operate an NC/CNC machine without first consulting the specific operator manual for the machine.
- Never place hands near a revolving spindle.
- Electrical cabinet doors are to be opened only by qualified personnel for maintenance.

Maintenance

A large investment has been made to purchase CNC equipment. It is very important to recognize the need for proper maintenance and a general upkeep of these machines. At the beginning of each opportunity to work on any Turning or Machining Center, verify that all lubrication reservoirs are properly filled with the correct oils. The recommended oils are listed in the operation or maintenance manuals typically provided with the equipment and sometimes a plate with a diagram of the machine and numbered locations for lubrication and the oil type is found on the machine. Most modern CNC machines have sensors that will not allow operation of the machine when the way or spindle oil levels are too low. Pneumatic (air) pressures need to be at a specified level and regulated properly.

If the pressure is too low, some machine functions will not operate until the pressure is restored to normal. The standard pressure setting is listed in Pounds per Square Inch (PSI) and a pressure regulator is typically located at the rear of the machine. Refer to the operator or maintenance manuals for recommended maintenance activities.

Coolant Reservoir

The Coolant tank should be cleaned and coolant level checked and adjusted as needed. A site glass is normally mounted on the tank. Use an acceptable water-soluble coolant mix or cutting oil. And last but not the least important is the cleanliness of the worktable, tools and area. Be sure to clear off any chips and remove any nicks or burrs on the clamping or mating surfaces. Always clean the machine after use.

Daily Maintenance Activities

Do's:

- Verify that all lubrication reservoirs are filled.
- Verify air pressure level by examining the regulator.
- Check the chip pan and coolant level and clean or fill, as needed.
- Make sure that automatic chip removal equipment is operational when the machine is cutting metal.
- Be sure that the worktable and all mating surfaces are clean and free from nicks or burrs.

Part 1   General Information

- ❏ Clean up the machine at the end of use with a wet/dry vacuum or wash machine guards with coolant to remove chips from the working envelope.

- ❏ Most new CNC machines are equipped with guards that envelope the worktable. The guards protect the ways and sensitive micro-switches installed as limit switches for table movement. Guards also help keep the surrounding floor space clean but there is still the task of chip disposal. Some larger production machines incorporate a chip conveyor, which carries the chips to a drum on the floor on either side of the machine.

Even with these features, there is still a need for chip cleanup inside the working envelope at least once a day. If chips are allowed to gather within the guards, they will eventually find their way around the guards that protect the machine ways. Over time, some of the chips might become embedded into the ways and cause irreparable damage. Another problem that may occur as the chips collect, they bunch up and are pushed into contact with the micro-switches. This contact stops the machine from working since the switches send a signal to the control that indicates table travel limit has been exceeded. This message prevents the machine from operating until the chips are removed. If chips get within the guards around the micro-switches, it is necessary to remove those guards and clean. Remember, it is essential to replace the guards after cleanup.

It is very important to do a thorough machine cleanup when many chips are present. The exterior of the machine usually will only need wiping down with a clean rag. Cleanup the ways and the working envelope without damaging the machine by using coolant to wash the machine table and the guards free of chips.

Another method is to use a wet/dry vacuum cleaner to pick up the chips. Along with the chip conveyor system, these two methods have proven hard to beat.

One cleanup method that is not recommended, is to use compressed air to blow away the chips. It is appropriate to use compressed air to remove chips and coolant from the workpiece itself or work holding fixtures such as a vise. The problem with using compressed air to clean up around the ways is that when chips are blasted away from the table, many are forced behind the guards, further worsening the micro-switch problem described above.

Tool Clamping Methods

The scope of this text is not intended to teach all of the necessary information regarding tooling. Consult the proper manuals for selection of cutting tools relevant to the required operation. Least expensive is not necessarily best. Proper selection of cutting tools and work holding methods are paramount to the success of any machining operation.

Always follow sound principals such as the most rigid set-up possible and do not allow large overhangs of tools or workpieces.

Just as with the rest of the machine tool, there are components used with the actual cutting tool that make it what it is. Obviously, the tool cutting edge is where the metal removal takes place. Without proper tool clamping, the cutting action may not produce the desired results. It is very important to carefully select the tool clamping method.

Part 1   General Information

In the case of a simple operation of milling a contour on a part, we may select a collet or a positive locking (posi-lock) holder for the end mill. The correct choice would depend on the actual features of the part to be machined. If the amount of metal to be removed is minimal, then a collet would probably suffice. If a considerable amount of metal is to be removed (more than two thirds of the tool diameter on a single depth of cut pass), the posi-lock holder selection is important. The reason for selecting the posi-lock holder is that under heavy cuts, a collet may not be able to grip the tool tightly enough. This situation could allow the tool to spin within the collet while in cut, with the result of ruining the collet and possibly damaging the part being machined. There is a tendency for the tool to dive into the workpiece when the tool spins within the collet. Note: Most High Speed Steel (HSS) end mills have a flat ground on them to facilitate the use of the posi-lock holder. This flat area allows for a set-screw to lock into it, creating a rigid and stable tool clamping method. The clamping method for drills could be either a collet or a drill chuck. A keyed drill chuck usually is used for heavier metal removal or larger holes, whereas the keyless type drill chuck is suitable for small holes. Generally, in the case of larger drills, a collet will be necessary to hold the tool. When drilling holes, remember to center-drill or spot-drill first, so that the tool does not walk. The center-drill may be held in the same manner as a drill.

In turning, the selection of the type of tool holder is determined by the finished part geometry and tools to be used. There are a variety of tool holder styles as well as insert shapes available to accomplish the part shape and size.

For more information on the proper selection of inserts and tool holders, refer to the Machinery's Handbook section titled "Indexable Inserts".

Another valuable resource for technical data regarding the selection of inserts and tool holding are the ordering catalogs from the tool and insert manufacturers.

Cutting Tool Selection

Tools are a very important aspect of machining. If the improper tool and/or tool clamping method is used the result will most likely be a poorly machined part. Always research the best tool and clamping method for a given operation. With the high speed and high performance of CNC machines, the proper selection process becomes increasingly important. The entire CNC machining process can be compromised by a lack of good tool planning.

There are many different machining operations to be performed on either turning or machining centers. The tool is where the action is, so if the improper selection takes place here, the whole machining sequence will be affected. Years of study have been dedicated to this subject and are documented within reference manuals and buyer's guides. Using these references will be helpful to correctly choose a tool for a given operation.

Remember that in your selection process, you are searching for the optimum metal-cutting conditions. The best way to understand how to choose the proper conditions is by studying the available data, such as the machine capabilities, the specific type of operation, the proper tool(s) and tool clamping method(s), the geometry of the part to be made,

Part 1 General Information

the workpiece and cutter material and the method of clamping the part.

It is important to utilize all of the advanced methods of metal removal available. Do not hesitate to research new technology. For example, in recent years, there have been numerous cutting tool innovations that include insert coatings (TIN), and insert materials (ceramic) to name a few. These advances have enabled increased cutting speeds and decreased tool wear providing for higher production throughput. Another innovation is modular tooling. This is a standardization of tool holders to facilitate the quick change of tools for setup and decreasing setup time. *Refer to the tool and insert ordering catalogs from the tool and insert manufacturers for more information on modular tooling.*

Tool Compensation Factors

Important information about the tool must to be given to the machine control unit (MCU), for the machine to be able to use the tool efficiently. In other words, the MCU needs the tool identification number, the tool length offset (TLO), and the diameter of each specific tool. A TLO is a measurement given to the control unit to compensate for the tool length when movements are commanded. The cutter diameter compensation (CDC) offset is used by the control to compensate for the diameter of the tool during commanded movements.

The tool number identifies where the tool is located within the storage magazine or turret and the sequence in which it is used. Each is assigned a tool length offset number. This number correlates with the pocket or turret position number and is where the measured offset distance from the cutting tip to the spindle face, in the case of a milling machine, is stored. For example, Tool No. 1 will have TLO No. 1. Finally, the diameter of the tool is compensated for. In most cases, the programmer has taken the diameter of the tool into account. In other words, the programmed tool path is written with a specific tool size in mind. However, more commonly the tool path centerline is programmed in order to facilitate the use of different tool diameters for a specified operation. When using the tool path centerline rather than the actual tool diameter, the additional offset is called cutter diameter compensation (CDC).

Tool Changing

CNC equipment enables more efficient machining by allowing the combination of several operations into a single setup. This combination of operations requires the use of multiple tools. Automatic tool changing (ATC) is a standard feature on most CNC Machining Centers, while many CNC Knee-Mills still require manual installation of the tool. Shown in the illustrations are two types of tool holders used on CNC machines; they have some distinct physical differences. Both of the holders are tapered. The one to the left has a single ring at the large end of the taper while the other has two rings. The tool holder that has only one ring is designed for machines that require manual tool

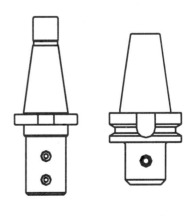

Figure 1   **Figure 2**

Part 1   General Information

changes. The tool with two rings is designed for machines that have automatic tool changers. These rings act as a gripping surface for a tool changer.

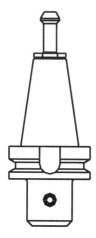

Figure 3

The tapered portion of the holder is the actual surface that is in contact with the mating taper of the spindle. These tapers are standardized by the industry and are numbered according to size (No. 30, No. 40, No. 50).

One benefit of these tapers over the standard R-8 Bridgeport style of tool holder is the increased surface area in contact with the mating taper of the spindle. The increased surface area makes the tool setup a more rigid and stable one.

Another feature on the tool holder is the notch or cutout on centerline of the tool (there is an identical cutout on the opposite side). This enables axial orientation within the spindle and tool changer. As the holder is inserted into the spindle, the cutouts enable it to be locked into place in exactly the same orientation each and every time it is used. This orientation makes a real difference when trying to perform very precise operations such as, boring a diameter. These notches also aid the spindle driving mechanism.

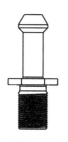

Figure 4

On CNC machines with a manual tool change, the holder is inserted into the machine and rotated until the holder pops into place (axial orientation is done by hand) and then, the draw bar is tightened to clamp the tool holder in place. Finally, another component of the CNC automatic tool changing system is the retention knob or pull-stud. Machining Centers need the retention knob/pull-stud to pull the tool into the spindle and clamp the holder. This knob is threaded into the small end of the taper as shown. Note: There are several styles of knobs available. The operator should consult the appropriate manufacturer manual for specifications required in their situation.

Metal Cutting Factors

Many tool and work holding methods used on manual machines are also used on CNC machines. The machines themselves differ in their method of control but otherwise they are very similar. The major objective of CNC is to increase productivity and improve quality by consistently controlling the machining operation. Knowledge of the exact capabilities of the machine and its components as well as the tooling involved is imperative when working with CNC. It is necessary for the CNC programmer to have a thorough knowledge of the CNC machines they are responsible for programming. This may involve an ongoing process of research with the goal of obtaining a near optimum metal-cutting process. From this research comes a decrease in the cycle time necessary to produce each part lowering per piece cost to the consumer. Fine tuning of the machining process for high-speed production gives more control over the quality of the product on a consistent basis.

Part 1 General Information

The Machine Tool

The machine used must have the physical ability to perform the machining. If the planned machining cut requires 10 horsepower from the spindle motor, a machine with only 5 horsepower will not be an efficient one to use. It is important to work within the capabilities of the machine tool. The stability and rigidity of the machine are of paramount importance as well.

The Cutting Fluid or Coolant

The metal cutting process is one that creates friction between the cutting tool and the workpiece. A cutting fluid or coolant is necessary to lubricate and remove heat and chips from the tool and workpiece during cutting. Water alone is not sufficient because it only cools and does not lubricate and it will also cause rust to develop. Also, because of the heat produced, water vaporizes and thus compromises the cooling effect. A mixture of lard-based soluble oil and water creates a good coolant for most light metal-cutting operations. Harder materials, like stainless steel and high alloy composition steels, require the use of a cutting-oil for the optimum results. Advancements have been made with synthetic coolants as well. Finally, the flow of coolant should as strong as possible and be directed at the cutting edge to accomplish its purpose. Operators should research, available resources like the *Machinery's Handbook*, for information about the proper selection of cutting fluids for specific types of materials.

The Workpiece & the Work Holding Method

The material to be machined has a definite effect on decisions about what tools will be used, the type of coolant necessary, and the selection of proper speeds and feeds for the metal-cutting operation.

The shape or geometry of the workpiece affects the metal-cutting operation and determines the type of work holding method that will be used. This clamping method is important for CNC work because of the high performance expected. It must hold the workpiece securely, be rigid, and not allow any flex or movement of the part at all.

The Cutting Speed

Cutting speed is the rate at which the tool moves past the workpiece in surface feet (sf/min) or meters per minute (m/min), to obtain satisfactory metal removal.

The cutting speed factor is most closely related to the tool life. Many years of research have been dedicated to this aspect of metal-cutting operations. The workpiece and the cutting tool material determine the recommended cutting speed. The *Machinery's Handbook* is an excellent source for information pertaining to determining proper cutting speed. If incorrect cutting speeds, spindle speed or feedrates are used the results will be poor tool life, poor surface finishes and even the possibility of damage to the tool and/or part.

The Spindle Speed

When referring to a milling or a turning operation, the spindle speed of the cutting tool must be accurately calculated relating to the conditions present. This speed is meas-

Part 1   General Information

ured in revolutions per minute, r/min (formerly known as RPM) and is dependant upon the type and condition of material being machined. This factor, coupled with a depth of cut, gives the information necessary to find the required horsepower necessary to perform a given operation. In order to create a highly productive machining operation all these factors should be given careful consideration. Refer to the formulae below needed to calculate r/min.

For Inch Units:

$$r/\min = \frac{12 \times CS}{\pi \times D}$$

For Metric Units:

$$r/\min = \frac{1000 \times CS}{\pi \times D}$$

where

CS = Cutting Speed from the charts in *Machinery's Handbook*

π = 3.1417

D = Diameter of the workpiece or the cutter.

Many modern machine controllers have a feature that allows automatic calculation of feeds and speeds that is based on operator input of the cutting conditions.

The Feedrate

Feedrate is defined as the distance the tool travels along a given axis in a set amount of time, generally measured in inches per minute in/min (formerly known as IPM) or inches per revolution in/rev (formerly known as IPR). This factor is dependent upon the selected tool type, the calculated spindle speed and the depth of cut. Refer to the Machinery's Handbook for the chip load recommendations and review the formula below that is necessary to calculate this aspect of the metal-cutting operation.

$$F = R \times N \times f$$

where

F = Feed in in/min or mm/min

R = r/min calculated from the preceding formula

N = the number of cutting edges

f = the chip load per tooth recommended from the *Machinery's Handbook*

The Depth of Cut

The depth of cut is determined by the amount of material to be removed from the workpiece, cutting tool flute length or insert size and the power available from the machine spindle. Always use the largest depth of cut possible to ensure the least affect on the tool life.

Cutting Speed, Spindle Speed, Feedrate and Depth of cut are all important factors in the metal-cutting process. When properly calculated, the optimum metal-cutting conditions will result.

Part 1   Process Planning for CNC

Process Planning for CNC

Certain steps must be followed in order to produce a machined part that meets specifications given on a blueprint. These steps need to be organized in a logical sequence to produce the finished part in the most efficient manner. Before machining begins, it is essential to go through the procedure called Process Planning. The following are the steps in the process:

1. Study the working drawing or blueprint.
2. Select the proper raw material or rough stock as described in the blue print.
3. Study the blueprint and determine the best sequence of individual operations needed to machine the required geometry.
4. Transfer the information onto planning charts.
5. Use in-process inspection to check dimensional values as they are completed while the part is still mounted on the machine.
6. Make necessary corrections and debar.
7. Perform a 100% dimensional inspection when the part is finished and log the results of the first article inspection on the quality control check sheet.
8. Take corrective action if any problems are identified.
9. Begin production.

Planning Documents

A blueprint may be thought of as a map it defines the destination. This destination is the end product. The roads available to get to this destination may be numerous. We do not start the trip without first determining what is the destination and how we are going to get there.

Planning charts resemble the required path to the destination. They are written descriptions of how to get there (to the end product). The following are descriptions of sample planning documents.

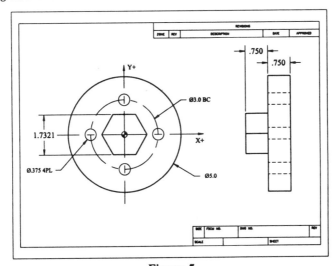

Figure 5

Part 1 Process Planning for CNC

The Blueprint

The information given on the blueprint will include the material, overall shape and the dimensions for part features. The geometry determines the type of machine (mill or lathe) to be used to produce the part. By studying the blueprint, material and operations (drilling, milling, boring, etc.) can be identified. The tools and work holding method can also be determined. Occasionally, the geometry will require multiple machines to manufacture the part, and additional operations will be necessary.

Operation Sheet

The purpose of this planning document is to identify the correct order for operations to be performed and the machine to be used. For example, suppose you are required to produce the part shown in Figure 5 above. Saw cutting the rough stock into blanks would come first. Then the part must be turned on a lathe to create the five-inch diameter and rough turn the diameter for the hexagon. Next a milling machine is used to cut the hexagon and drill the bolthole circle. Before any inspection can be carried out, the part must be deburred. Finally the part can be inspected for accuracy.

YOUR COMPANY
OPERATION SHEET

Date:	Prepared By:		
Part Name:	Part Number:		
Quantity:	Sheet____ of____		
Material:			
Raw Stock Size:			
Operation Number	Machine Used	Description of Operation	Time

Chart 1

Part 1   Process Planning for CNC

The operation sheet is particularly useful when many identical parts are machined (production run). The operation sheet is similar to directions or a how-to approach. The process needed to manufacture the finished part has been decided in advance and is documented for use.

When small batches of parts are to be made, there may not be an operation sheet. It is the machinist's responsibility to study the blueprint and decide the necessary steps to machine the part. The operation sheet can aid in this decision making process.

With CNC machining, multiple part geometry features can be performed in one setup. In some cases, when using a CNC Machining Center, a part might be machined to its completed status without ever using another machine. This is very efficient and another advantage of the use of CNC equipment.

To complete an operation sheet, study the blueprint; then decide on the steps necessary to machine the part. Document the machining process and refine any problems the process has, then list the operations in the correct sequence in which they will be performed.

The top section of the operation sheet is for reference information and includes:

The date the document is prepared or revised

The name of the person preparing it

The part name and the part number (from the blueprint)

The quantity of parts to be manufactured

Since some parts may require a large number of operations, it is possible that more than one operation sheet will be needed to document the whole process. The top section also includes a sheet numbering system (Sheet _____ of _____). This information must be included. Other information included on the operation sheet header is the material, the raw stock size for the part, followed by the operations list.

CNC Setup Sheet

The CNC Setup Sheet has two sections. The top section is for reference information and includes:

The date the document is prepared or revised

The name of the person preparing it

The part name and part number (from the blueprint)

The machine being used

Note: If more than one machine is to be used to manufacture a single part, separate setup sheets are completed for each machine.

The CNC part program used in the manufacturing process

Workpiece Zero reference points for the part (program zero)

Work holding devices

Note: If more than one device is needed, the operation number(s) and process are also included.

Part 1   Process Planning for CNC

The lower half of this form lists the tool(s) by number, description and offset. There is a column for comments, remarks or explanations, if needed.

YOUR COMPANY
CNC SETUP SHEET

Date:	Prepared By:
Part Name:	Part Number
Machine:	Program Number:

Workpiece Zero X _____ Y _____ Z _____

Setup Description:

Tool #	Description	Offset	Comments

Chart 2

Quality Control Check Sheet

This planning document is used for the final inspection stage of the machining process. Once the part is completed, it is necessary to check all of the dimensions listed on the blueprint to verify they are within the specified tolerance. The Quality Control Check Sheet is an excellent method to document the results of this inspection and a valuable tracking tool.

Reference information is similar to the other planning documents. Included are:

The date the document is prepared or revised

The name of the person checking the part

The part name and part number (from the blueprint).

On the check sheet, 100 % of the blueprint dimensions and their tolerances are written down in list form. Using this method, sequentially go through the each of the print dimensions and log the results. This assures that the machined part meets the specifications given on the blueprint. If dimensions are found that do not meet specifications, corrective action must be taken.

Part 1 Process Planning for CNC

YOUR COMPANY
QUALITY CONTROL CHECK SHEET

Date:	Part Name:	Part Number:	Checked By:
Blueprint Dimension	Tolerance	Actual Dimension	Comments

Chart 3

Types of Numerically Controlled Machines

There are two basic groups of numerically controlled machines, Numerical Control (NC) and Computer Numerical Control (CNC).

In an NC system, the program is run from a punched tape; it is impossible to store such a program in memory. For a punched tape to be used again to machine another part, it must be rewound and read from the beginning. This routine is repeated every time the program is executed. If there are errors in the program and changes are necessary, the tape will need to be discarded and a new one punched. The process is costly and error prone and while this type is still in use, it is becoming obsolete.

Machines with a CNC system are equipped with a computer, consisting of one or more microprocessors and storage facilities. Some have hard drives and are network configurable. Program data are entered through the control panel keyboard or via an RS232 communications port from a remote source like a Personal Computer (PC). The control panel enables the operator to make corrections (edits) in the program stored in memory, thereby eliminating the need for new punched tape.

Types of CNC Machines have expanded vastly over the last decade. Turning and Machining Centers are the focus of this book but there are many other types of machines using Computerized Numerical Control. For example there are: Electrical Discharge Machines (EDM), Grinders, Lasers, Turret Punches, and many more. Also, there are many different designs of Machining and Turning Centers. Some of the Machining

Part 1 Process Planning for CNC

Centers have rotary axes and some Turning Centers have live tooling and secondary spindles. For this text the focus will be limited to Vertical Machining Centers with three axes and Turning Centers with two axes. These types of machines are considered the foundation of all CNC learning. All operations on these machines can be carried out automatically. Human involvement is limited to loading and unloading the workpiece and entering the amounts of dimensional offsets.

What is Programming?

Programming is, a method of defining tool movements through the application of numbers and corresponding coded letter symbols. As shown in the list below, all phases of production are considered in programming, beginning with the technical part drawing and ending with the final product:

Technical Part Drawing

Work Holding Considerations

Tool Selection

Preparation of the Part Program

Part Program Tool Path Verification

Measuring of Tool and Work Offsets

Program Test by Dry Run

Automatic Operation or CNC Machining

All programming begins by a close evaluation of the technical drawing and emphasizes assigned tolerances for particular operations, tool selection, and the choice of a machine. The next step is the selection of the machining process. The machining process refers to the selection of fixtures and determination of the operation sequence. Following that, is a selection of the appropriate tools and determination of the sequence of their application. Before writing a program, spindle speeds and feed rates have to be calculated.

When program writing begins, special attention is given to the specific tool movements necessary to complete the finished part geometry, including non-cutting movements. Individual tools are identified and noted in the manuscript. Also, the miscellaneous and feeds and speeds functions necessary are noted for each tool such as; flood coolant, spindle direction, r/min and feedrates (these items will be covered in greater detail in the following chapters). Then, once the program is written it must be transferred to the machine through an input medium like one of the following: punched tape, floppy disk, or by RS232 interface.

Machining is initiated by preparing the machine for use, commonly called setup (for example, input of workpiece zero and tool length offset into CNC memory). Many modern controllers have a function for graphical simulation of the programmed tool path on the Cathode Ray Tube (CRT). This enables the machinist or set-up person to verify that the program has no errors, and to visually inspect tool movements. If all looks well, the first part can be machined with increased confidence. After completion, a thorough inspection will compare dimensions of the final product to those on the drawing. Any differences between the actual dimensions and the dimensions on the drawing are inserted into the offset register of the machine. In this manner, the correct dimensions of consecutively machined parts can be obtained.

Part 1   Coordinate Systems

Orientation of the Coordinate System

All machines are equipped with the basic traveling components, which move in relation to one another as well as in perpendicular directions. Turning Centers are equipped with a tool carrier, which travels along two axes. (Figure 6 and Figure 7)

Note that in the following drawings of lathes the tool is located on the positive side of spindle centerline. This is a common design of modern CNC lathes. For visualization purposes in this book the cutting tool will be shown upright. In reality it is mounted with the insert facing down and the spindle is rotated clockwise for cutting.

Note: the direction of spindle rotation, clockwise (CW) or Counterclockwise (CCW), in turning, is determined by looking from the headstock towards the tailstock and tool orientation.

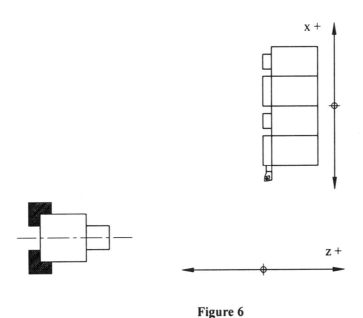

Figure 6

Machining Centers are milling machines are equipped with a traversing worktable, which travels along two axes, and a spindle, whose tool travels along a third axis. (see following page Figure 8)

All axes of machines are oriented in an orthogonal (each axis is perpendicular to the other) coordinate system, for example, Cartesian Coordinate System (right-hand rule system). (see following page Figure 9)

Part 1   Coordinate Systems

Figure 7

Figure 8

The Right-Hand Rule System

In discussing the X, Y, and Z axes, the right hand rule establishes the orientation and the description of tool motions along a positive or negative direction for each axis. This rule is recognized worldwide and is the standard for which axis identification was established.

Use Figure 9 on the following page to help visualize this concept. For the vertical representation if the palm of your right hand is laid out flat in front, the thumb will point in the positive X direction. The forefinger will be pointing the positive Y direction. Now fold over the little finger and the ring finger and allow the middle finger to point

Part 1   Coordinate Systems

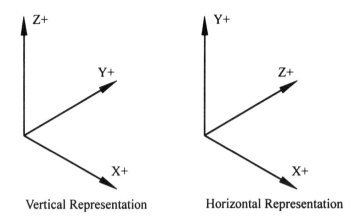

Vertical Representation   Horizontal Representation

Figure 9

up. This forms the third axis, Z, and points in the Z positive direction. The point where all three of these axes intersect is called the origin or zero point. When looking at any vertical milling machine, you can apply this rule.

Coordinate Systems

Visualize a grid on a sheet of paper (graph paper) with each segment of the grid having a specific value. Now place two solid lines through the exact center of the grid and perpendicular to each other. By doing this you have constructed a simple two-dimensional coordinate system. Carry the thought a little further and add a third imaginary line. This line passes through the same center point as the first two lines but as vertical; that is, it rises above and below the sheet on which the grid is placed. This additional line would represent the third axis in the three dimensional coordinate system which is called the Z axis.

Two-Dimensional Coordinate System

A two-dimensional coordinate system, such as the one used on a lathe, only uses the X and Z axes for measurement. The X runs perpendicular to the workpiece and the Z

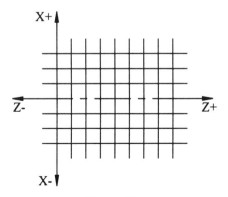

Figure 10

19

Part 1   Coordinate Systems

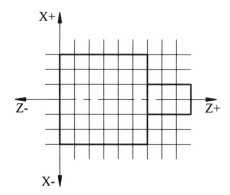

Figure 11

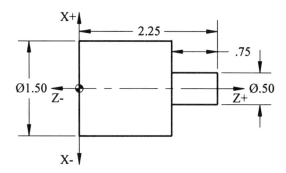

Figure 12

axis is parallel with the spindle centerline. When working on the lathe, we are working with a workpiece that has only two dimensions, the diameter and the length. On blueprints, the front view generally shows the features that define the finished shape of the part for turning. In order to see how to apply this type of coordinate system, study the above diagrams.

Think of the cylindrical work piece as if it were flat or as shown in the front view of the blueprint. Next, visualize the coordinate system superimposed over the blueprint of the workpiece, aligning the X axis with the centerline of the diameter shown. Then align the Z axis with the end of the part, which will be used as an origin or zero-point. In most cases, the finished part surface nearest the spindle face will represent this Z axis datum and the centerline will represent the X axis. Where the two axes intersect is the origin or zero point. By laying out this "grid," we now can apply the coordinate system and define where the points are located to enable creation of the geometry from the blueprint. Another point to consider: on a lathe, the cutting takes place on only one side of the part or the radius, because the part rotates and it is symmetrical about the centerline. In order to apply the coordinate system, all we need is the basic contour features of one-half of the part (on one side of the diameter). The other half is a mirror image; when given this program information, the lathe will automatically produce the mirror image.

Part 1 Coordinate Systems

Three Dimensional Coordinate System

Although the mill is three dimensional, the same concept (using the front view of the blueprint) can be used with rectangular workpieces. As with the lathe, the Z axis is related to the spindle. However, in case of the three dimensional rectangular workpiece, the origin or zero-point must be defined differently. In the example shown in Figure 13, the lower left hand corner of the workpiece is chosen as the zero-point for defining movements using the coordinate system. The thickness of the part is the third dimension or Z axis. When selecting a zero-point for the Z axis of a particular part, it is common to use the top surface.

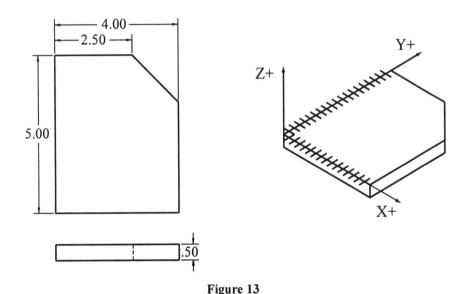

Figure 13

The Polar Coordinate System

If a circle is drawn on a piece of graph paper so that the center of the circle is at the intersection of two lines and the edges of the circle are tangent to any line on the paper this will help in visualizing the following statements. Let's consider the circle center as the origin or zero-point of the coordinate system. This means that some of the points defined within this grid will be negative numbers. Now draw a horizontal line through the center and passing through each side of the circle. Then draw a vertical line through the center also passing through each side of the circle. Basically, we've made a pie with four pieces. Each of the four pieces, or segments of the circle is known as a quadrant. The quadrants are numbered and progress counter-clockwise. In Quadrant No. 1, both the X and Y axis point values are positive. In Quadrant No. 2, the X axis point values are negative and the Y axis point values are positive. In Quadrant No. 3, both the X and Y axis point values are negative. Finally, in Quadrant No. 4, the X axis point values are positive while the Y axis values are negative. This quadrant system is true regardless of the axis that rotation is about. The following drawings illustrate the values (negative or positive) of the coordinates, depending on the quarter circle (quadrant) in which they appear.

Part 1 Coordinate Systems

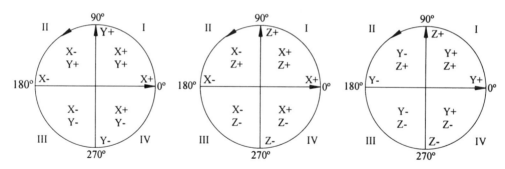

Figure 14

Although the rectangular coordinate system can be used to define points on the circle, a method using angular values may also be specified. We still use the same origin or zero-point for the X and Y axes. However, the two values that are being considered are an angular value for the position of a point on the circle and the length of the radius joining that point with the center of the circle. To understand the polar coordinate system, imagine that the radius is a line circling around the center origin or zero-point. Thinking in terms of hand movements on a clock, the three-o'clock position has an angular value of 0° counted as the "starting point" for the radius line. The twelve-o'clock position is referred to as the 90° position, nine-o'clock is 180° and the six-o'clock position is 270°. When the radius line lies on the X axis in the three-o'clock position, we have at least two possible angular measurements. If the radius line has not moved from its starting point, the angular measurement is known as 0°. On the other hand, if the radius line has circled once around the zero point, the angular measurement is known as 360° Therefore, the movement of the radius determines the angular measurement. If the direction in which the radius rotates is counter-clockwise, angular values will be positive. A negative angular value (such as -90°) indicates that the radius has rotated in a clockwise direction. Note: A 90° angle (clockwise rotation) places the radius at the same position on the grid as a +270° (counter-clockwise) rotation.

Sometimes the blueprint will not specify a rectangular coordinate but will give a polar system in the form of an angle. With some basic trigonometric calculations, this information can be converted to the rectangular coordinate system.

The same polar coordinates system applies regardless of the axis of rotation as is shown in Figure 14. When rotation around the X (A), Y (B), or Z (C), are encountered, this is considered an additional axis known as the fourth axis.

All operations of CNC machines are based on three axes: X, Y, and Z.

1. (X0, Y0, Z0)
2. (X0, Y0, Z+)
3. (X0, Y-, Z+)
4. (X0, Y-, Z0)
5. (X-, Y-, Z0)

Part 1 Coordinate Systems

6. (X-, Y0, Z0)
7. (X-, Y0, Z+)
8. (X-, Y- Z+)

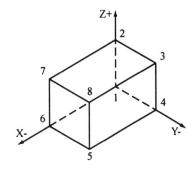

Figure 15

Figure 15 illustrates a box-like object in which one vertex (point 1) is located at the origin of the coordinate system. We have defined the location of all box vertices by specifying coordinate signs for all their points above. Note the position of the coordinate system on the following machines.

On vertical milling machines the spindle axis is perpendicular to the surface of the worktable.

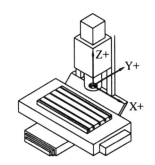

Figure 16

On horizontal milling machines the spindle axis parallel to the surface of the worktable.

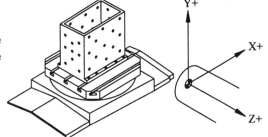

Figure 17

On lathes, the spindle axis is also the workpiece axis.

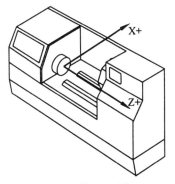

Figure 18

Part 1   Coordinate Systems

Points of Reference

When using CNC machines, any tool location is controlled within the coordinate system. The accuracy of this positional information is established by specific Zero Points (reference points). The first is Machine Zero, a fixed point established by the manufacturer that is the basis for all coordinate system measurements. On a typical lathe, this is usually the spindle centerline in the X axis and the face of the spindle nose for the Z axis. For a milling machine, this position is often at the furthest end of travel in all three axes in the positive direction. Occasionally, this X axis position is at the center of the table travel.

This Machine Zero Point establishes the coordinate system for operation of the machine and is commonly called Machine Home (Home Position) . Upon startup of the machine, all axes needs to be set to this position first. The Machine Zero Point identifies to the machine controller where the origin for each axis is located.

The Operator's manual supplied with the machine should be consulted to identify where this location is and how to properly Home the machine.

The second zero point can be located anywhere within the machine work envelope and is called Workpiece Zero that is used as the basis for programmed coordinate values used to produce the workpiece. It is established within the part program by a special code and is dependant on the Machine Zero point. The code in the program identifies the location of offset values to the machine control where the exact coordinate distance of the X, Y and Z axes of Workpiece Zero is in relationship to the Machine Zero. All dimensional data on the part will be established by accurately setting the Workpiece Zero. A way of looking at the Workpiece Zero is like another coordinate system within the machine coordinate system, established by the Home Position.

Tool offsets are also considered to be Zero Points and are compensated for with Tool Length and Diameter Offsets. The tool setting point for a lathe has two dimensions; the distance on diameter from the tool tip to the centerline of the tool turret, and the distance from the tool turret face to the tool tip. The tool setting point for the mill is the distance from the spindle face to the tool tip, and the distance from the tool tip to the spindle centerline.

Blueprint Relationship to CNC

The standard called ASME Y14.5-1994 establishes a method for communicating part dimensional values in a uniform way on the engineering drawing or blueprint. The drawing information will be translated to the coordinate system in order for dimensional values and part features to be manufactured.

On the blueprint, Datum features are identified as Primary (A) Secondary (B) and Tertiary (C). Dimensions for the workpiece are derived from these datum features. On the drawing, the point where these three datum features meet is called the origin or zero point for the part. When possible, this same point should be used for Workpiece Zero. This allows the use of actual blueprint dimensions within the part program and often results in fewer calculations. Most drawings are developed using an absolute dimensioning system based on Datum dimensions derived from the same fixed point (origin or zero point). Occasionally, some features may be dimensioned from the location of anoth-

Part 1   Coordinate Systems

er feature. An example of this might be a row of holes exactly one half of an inch apart. This type of dimensioning is called relative or incremental.

Note: A thorough knowledge of blueprint reading is required for successful results using manual or CNC equipment.

Machine Zero

Each CNC machine is assigned a fixed point, which is referred to as Machine Zero (or Machine Home). For most machines, Machine Zero is defined as the extreme end position of main machine components that are oriented in a given coordinate system. From Machine Zero, we can determine the values of the coordinates that, in turn, determine the position of the point commanded in a CNC program. Electromechanical sensors called micro-switches (limit switches) are located in the extreme end positions of traveling machine components. These sensors send a signal to the controller when they are activated and thus setting the home position. In the case of milling machines, Machine Zero on the table is set with respect to the X and Y axes. Position zero on the spindle is set with respect to the Z axis. Machine Zero of the tool carrier on lathes is set with respect to the X and Z axes. Positioning the traveling components at zero can be performed manually as well as with the use of the control panel or directly in the program, employing a Machine Zero function. After turning the machine power on it is required to position the machine components at Machine Zero before proceeding further. From that point on, all machine components will always automatically return to the same exact position when required.

Machine Zero is frequently the position in which tool changes take place. Therefore, if you intend to change the tool before a given operation, then the machine must be positioned at Machine Zero for the Z axis on vertical machines and the Y axis on horizontal machines.

Workpiece Zero

So far, for all main traveling components of numerically controlled machines, we have assigned an oriented axis within the coordinate system. Any movement of machine components must be described by points, which actually determine the traveling path of the tool. Changes in the position tool are determined with respect to the stationary reference point of machine zero.

In order to better understand this concept, this situation can be illustrated with a rectangular plate in which all coordinates are described at their four corners (P_1, P_2, P_3, P_4).

P_1 = X -15.0, Y-10.0

P_2 = X -15.0, Y-12.0

P_3 = X -20.0, Y-12.0

P_4 = X -20.0, Y-10.0

Part 1 Coordinate Systems

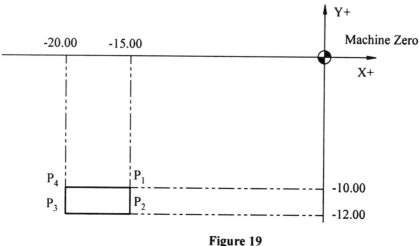

Figure 19

Determine the coordinates of these points. The rectangle has been placed in such a manner that each side is parallel to one axis of the coordinate system. If the distance from Machine Zero is measured to any point on the workpiece, the coordinates of the remaining points can be determined from the dimensions given on the drawing.

All programmed point coordinates, whose values are determined with respect to Machine Zero, must be given every time which is time consuming. It may also cause errors due to the fact that all the given dimensions determining the points do not always refer to those on the drawing. As previously mentioned, in order to determine the coordinates for the four corners of the rectangular part illustrated, it is necessary to find the distance between machine zero and a specific point of reference. Then, all the remaining dimensional data to be used are taken from the blueprint.

For all numerically controlled machines, we follow certain principles to define the method of selecting Workpiece Zero. At the beginning of the program, we input the value of the distance between machine zero and the selected Workpiece Zero, by employing function G92 or G54 through 59 for Machining Centers and function G50 for Turning Centers. Let us review the same situation and note the changes of the point coordinates when applying Workpiece Zero.

G92 x 15.0Y10.0 or G54-59 X 0.0Y0.0

P_1 = X0, Y0

P_2 = X0, Y-2.0

P_3 = X-5.0, Y-2.0

P_4 = X-5.0, Y0

The values for G92 or G54 through 59 are valid until they are recalled by the same function, but with different coordinates, for X and Y. When programming Machining Centers, we place function G92 or G54 through 59 only at the beginning of the program, whereas the values assigned to function G50 for Turning Centers will need to be added for each tools position. Once this activation is read by the

Part 1 Coordinate Systems

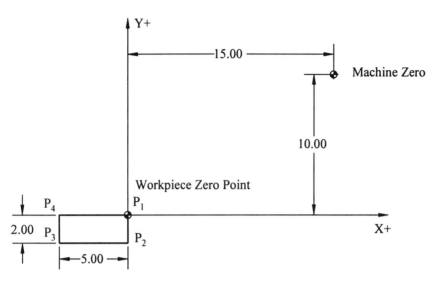

Figure 20

control, all coordinates will be measured from the new Workpiece Zero allowing the use of part dimensions for programmed moves.

With Turning Centers, Workpiece Zero in the direction of the Z axis is on the face surface of the workpiece, and the centerline axis of the spindle is Workpiece Zero in the direction of the X axis.

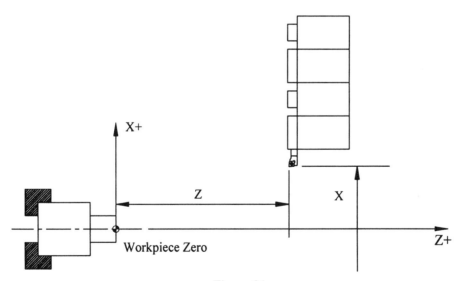

Figure 21

On Machining Centers, Workpiece Zero is frequently located on the corner of the workpiece or in alignment with the Datum of the workpiece.

The application of Workpiece Zero is quite advantageous to the programmer because the input values of X, Y, and Z in the program can be taken directly from the drawing. If the program is used another time, the values of coordinates X and Y

Part 1   Coordinate Systems

(assigned to functions G50 and G92 or G54 through G59) will have to be inserted again, prior to automatic operation.

Absolute and Incremental Coordinate Systems

When programming in an absolute coordinate system, the positions of all the coordinates are based upon a fixed point or origin of the coordinate system. The tool path from point P_1 to P_{10}, for example, is illustrated below.

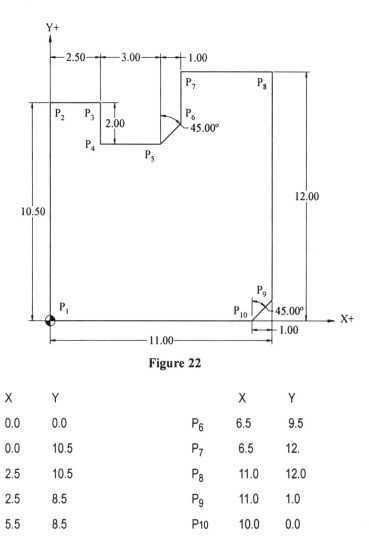

Figure 22

	X	Y		X	Y
P_1	0.0	0.0	P_6	6.5	9.5
P_2	0.0	10.5	P_7	6.5	12.
P_3	2.5	10.5	P_8	11.0	12.0
P_4	2.5	8.5	P_9	11.0	1.0
P_5	5.5	8.5	P_{10}	10.0	0.0

Programming an incremental coordinate system is based upon the determination of the tool path from its current position to its next consecutive position and in the direction of all the axes. Sign determines the direction of motion. Based on the drawing from the previous example, we can illustrate the tool path in an incremental coordinate system, starting and ending at P_1.

Part 1 Coordinate Systems

	X	Y		X	Y
P_2	0.0	10.5	P_7	0.0	2.5
P_3	2.5	0.0	P_8	4.5	0.0
P_4	0.0	-2.0	P_9	0.0	-11.0
P_5	3.0	0.0	P_{10}	-1.0	-1.0
P_6	1.0	1.0	P_1	-10.0	0.0

Input Format

Numerically controlled machines allow input values of inches specified by the command G20, millimeters specified by the command G21, and degrees with a decimal point and significant zeros in front of (leading) or at the end (trailing) of the values. When using inch programming the two ways distances can be specified:

Programming with a decimal point
    1 inch =1. or 1.0
    1 1/4 inch =1.250 or 1.25
    1/16 inch =0.0625 or .0625

Programming with significant trailing zeros

In this case, the zero furthest to the right corresponds with the ten thousandths of an inch.
    1 inch = 10000
    1 1/8 inch = 11250
    1 1/32 inch = 10313

These two methods are the standard on all CNC machines

With modern controllers neither leading nor trailing zeros are required, the decimal placement is the significant factor. In this case the input is as follows:
    1 inch =1. or 1.0
    1 1/4 inch =1.25
    1 1/16 inch =1.0625

Program Format

The language described in this book is used for controlling machine tools is known informally as "G-Code". This language is used worldwide and is reasonably consistent. A program created for a particular part on one machine may be used on other machines with minimal changes required.

The following chart is a list for all of the addresses applicable in programming, along with brief explanations for each:

Part 1   Program Format

Address Characters

Character	Meaning
A	Additional rotary axis parallel and around the X axis
B	Additional rotary axis parallel and around the Y axis
C	Additional rotary axis parallel and around the Z axis
D	Tool radius offset number Depth of cut for multiple repetitive cycles
E	User macro character Precise designation of thread lead
F	Feed rate Precise designation of thread lead
G	Preparatory function
H	Tool length offset number
I	Incremental X coordinate of circle center or parameter of fixed cycle
J	Incremental Y coordinate of circle center
K	Incremental Z coordinate of circle center or parameter of fixed cycle
L	Number of repetitions
M	Miscellaneous function
N	Sequence or block number
O	Program number
P	Dwell time, program number, and sequence number designation in subprogram Sequence number for multiple repetitive cycles
Q	Depth of cut, shift of canned cycles Sequence number for multiple repetitive cycles
R	Point R for canned cycles, as a reference return value Radius designation of a circle arc
S	Spindle-speed function
T	Tool function
U	Additional linear axis parallel to X axis
V	Additional linear axis parallel to Y axis
W	Additional linear axis parallel to Z axis
X	X coordinate
Y	Y coordinate
Z	Z coordinate

Chart 4

Part 1   Address Characters

Each program is a set of instructions that controls the tool path. The program is made up from blocks of information separated by the semicolon symbol (;). This symbol (;) is defined as the end of the block (EOB) character. Each block contains one or more program words. For example:

Word	Word	Word	Word	Word
N02	G01	X3.5	Y4.728	F8.0

Each word contains an address, followed by specific data. For example:

Address	Data	Address	Data	Address	Data
N	02	G	01	X	3.5

Part I
Study Questions

1. Programming is a method of defining tool movements through the application of numbers and corresponding coded letter symbols.
   T or F

2. A lathe has the following axes:
   a. X, Y & Z   c. X & Z only
   b. X & Y only   d. Y & Z only

3. Program coordinates that are based on a fixed origin are called:
   a. Incremental   c. Relative
   b. Absolute   d. Polar

4. On a two axis turning center the diameter controlling axis is:
   a. B   b. A   c. X   d. Z

31

Part 1 Study Questions

5. The letter addresses used to identify axes of rotation are:
   a. U, V & W
   c. A, Z & X
   b. X, Y & Z
   d. A, B & C

6. The acronym TLO stands for:
   a. Tool Length Offsets
   c. Taper Length Offset
   b. Total Length Offset
   d. Time Length Offset

7. When referring to the polar coordinate system, the clockwise rotation direction has a positive value
   T or F

8. In Figure 15 of Part I, which quadrant is the part placed in?

9. A program block is a single line of code followed by an end-of-block character
   T or F

10. Each block contains one or more program words.
    T or F

11. Using Figure 13 in Part I, list the X and Y absolute coordinates for the part profile where workpiece zero is at the lower left corner. (The corner cutoff is at a 45° angle).

12. Using Figure 13 in Part I, list the X and Y incremental coordinates for the part profile where workpiece zero is at the lower left corner.

13. How often should the machine lubrication levels be checked?

PART 2
Operation

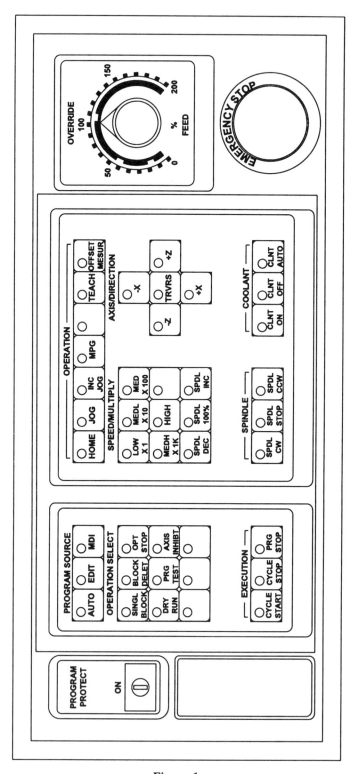

Figure 1

Operation Panel Descriptions

Figure 2

The following descriptions for the above diagram represent the configuration for a common Operators Panel. Some differences do exist for each manufacturer Operation Panels but they generally contain the same features. The illustration shows a panel for a two-axis lathe. The panel used for a mill would be essentially identical except for the added keys for the additional axes. The user should consult the applicable manufacturer manual for detailed descriptions that match their needs. Please also note that the Handle (Pulse Generator) and the Rapid Traverse Override buttons are not shown in this view of the controller although it is described in the text.

Another common item not shown here are the switches used to change the chucking direction from external to internal and are specifically for lathes.

Operator Panel Features

Feedrate Override

FEEDRATE OVERRIDE, This switch allows control of the feedrate when the operator adjusts the position. At the 12:00 o'clock position the feed, during autocycle, will occur at 100% of the programmed value. This allows the control of the work feeds defined by the F-word in the program. The percentage of the value entered in the program can be increased or decreased.

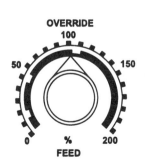

Figure 3

Part 2   Operation Panel Descriptions

This feature offers the operator the control needed to fine tune feeds. It can also be used to control feedrates during jog mode function.

Emergency Stop

Figure 4

The EMERGENCY STOP is the large, red, mushroom-shaped button used to stop machine function when an emergency situation occurs. Some example situations are: overloading of the machine, a machined part is loose, or incorrect data in the program or work tool offsets are causing a collision (crash) between the tool and the workpiece. When this button is depressed, all program commanded feedrates and spindle revolutions are halted immediately. To recover from an "E-Stop" condition requires resetting the program controller and Homing of the machine axes. To reset the EMERGENCY STOP button, turn it clockwise. It should "pop-out" of the depressed condition. Check the monitor for any alarm signals and take note of the Alarm # and description, then eliminate the cause that forced the use of the EMERGENCY STOP button. Press the Reset button to clear all pending commands and Home the machine axes when no interference conditions are present.

Program Protect

Figure 5

When this key-switch is in the ON condition (vertical), it prohibits any program changes to be made. The condition does not affect work or tool offset adjustments. Some shops set this condition to ON, remove this key and allow only the programmer or set-up person access to the key. This is especially true in larger shops with multiple shifts and many people.

Program Source

Figure 6

On some operator panels, a rotary switch referred to as Mode Select is used instead of buttons shown here. This switch includes both automatic (AUTO) and manual operation functions. The position of this switch determines whether the machine utilizes the automatic or the manual control. This switch can also be positioned to allow the entry of data into the control manually (Manual Data Input or MDI) or to make changes to the program through the EDIT mode. For this purpose, operator panel buttons are used to specify the control or operational mode.

Note: When the buttons are pressed, they are active and remain so until another mode button that overrides it is pressed. In some conditions, multiple Light Emitting Diodes (LED's) may be lit simultaneously. The LED in the upper left corner of the button is lit when the mode is ON and active.

Part 2   Operation Panel Descriptions

Auto

By pressing this button, the control mode enables the CNC commands stored in the memory to be executed for automatic operation.

Edit

By pressing this button, the program edit mode is selected. The EDIT mode enables you to enter the program to control memory, enter any changes to the program and transfer data from the program via RS232 interface to an offline storage device or check the program file memory and storage capacity.

Note: The RS232 interface is a 25-pin serial cable connector (shown in the illustration) located behind the flip-up door just below the Program Protect key switch. This connection is also used for DNC (Direct Numerical Control) program operations when the program is too large for the controller memory.

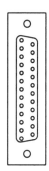

Figure 7

MDI-Manual Data Input

By pressing this button, the MANUAL DATA INPUT mode is selected.

The MDI mode enables the automatic control of the machine, using information entered in the form of program blocks without interfering with the basic part program. This mode is often used during the machining of workpiece holding equipment such as soft-jaws and during setup. It corresponds to single moves (milling surfaces, drilling holes), descriptions of which need not be entered to memory storage. MDI mode can also be used during the execution of the program. For example, suppose the program is missing the command M03 S350; needed to turn on the spindle, clockwise, at 350 RPM and the End-Of-Block character (;). In order to correct this omission, press the SINGLE BLOCK button and then the MDI button. Using MDI, you can enter functions M03 and S350 from the control panel keypad.

Enter this command by pressing INPUT on the control panel. Press AUTO to reenter the program auto-cycle mode and CYCLE START to continue execution of the program from memory.

Operation Select

The following buttons are related to the automatic operation of the machine. Activating one of these has an effect on the operation as described below:

Single Block

The execution of a SINGLE BLOCK (SINGL BLOCK) of information is initiated by pressing this button to turn it ON. Each time the CYCLE START button is pressed, only one block of information will be executed. This switch can also be used if you intend to check the performance of a new program on the machine or when the momentary interruption of a machine's work is necessary.

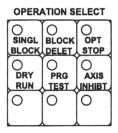

Figure 8

Part 2 Operation Panel Descriptions

Block Delete

BLOCK DELETE (BLOCK DELET) is sometimes called Block Skip. When this button is pressed before the auto-cycle mode is entered, the controller skips execution of the program blocks that start with "/"and end with the end of block (;) character. For instance, if part of the program, or a particular block of the program is not presently needed, but you would like to keep this information for future use, then place the block skip symbol "/" at the beginning of each such block. The BLOCK DELET button is located on the control panel. If it is turned on, information contained in the blocks that contain the symbol "/" will not be executed.

Example:

N100G01X2.810Y3.256

/N01X3.253Y2.864

/N02X3.800

(Blocks N01 and N02 will be skipped.)

Notes: The symbol "/" should be placed at the beginning of the block. If it is not, then all the information contained in the block preceding the symbol "/" will be executed, while the information following this symbol will be omitted.

If the BLOCK DELET is in the OFF condition, all blocks (regardless of the symbol "/") will be executed.

When transferring the program to punched tape or external computer, all the information (regardless of symbol "/") is transferred.

Opt Stop-Optional Stop

When this button is pressed, the OPTIONAL STOP mode is active.

The OPTIONAL STOP interrupts the automatic cycle of the machine if the program word M01 appears in the program. Quite often, function M01 is placed in the program after the work of a particular tool is completed or before a tool change. This enables the operator to perform a routine measurement directly on the machine and, if necessary, make adjustments and then rerun the same tool.

Dry Run

By pressing the DRY RUN button during automatic cycle, all of the rapid and work feeds are changed to the rapid traverse feed set in the parameters instead of the programmed feed. Consult the manufacturer manual for specific directions on the use of this function.

DRY RUN is also used to check a new program on the machine without any work actually being performed by the tool. This is particularly useful on programs with long cycle times so the operator can progress through the program more quickly.

Use caution when using this function, it is not intended for metal cutting.

Part 2   Operation Panel Descriptions

Prg Test-Program Test

This function is also known as MACHINE LOCK. Activating this mode inhibits axis movement on all of the axes. This button is used to check a new program on the machine through the controller. All movements of the tool are locked, while a program check is run on the computer and displayed on the screen. The operator can observe the position display on the screen. If any program errors are encountered, an alarm will be displayed. This function is used to check a program and is especially useful for very large programs requiring a long cycle time to complete. This test is normally the first in a series (Program Test, Dry Run and Single Block) of preliminary actions to be executed before full auto cycle mode is attempted. For any program test, all offsets should be set first.

Axis Inhbt-Axis Inhibit

This function is identical to MACHINE LOCK for all axes. Activating this mode inhibits axis movement on all of the axes. A common situation would be to inhibit the axes to allow for internal checking of the program. Some controllers have additional buttons or switches that enable inhibiting of only one axis at a time. This function is especially useful when inhibiting the Z axis so that all X, Y movements can still be observed

Execution

These three buttons also have to do with automatic operation of the machine. The first button starts, the second temporarily stops automatic operation and the last key merely indicates when a program stop is encountered. Their specific functions are described below:

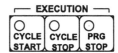

Figure 9

Cycle Start

The CYCLE START button is used to start automatic operation.

Use this button in order to begin the execution of a program from memory. When the CYCLE START button is pressed, the control lamp located above this button goes on and the active program will be executed to the end.

Cycle Stop (Feed Hold)

Pressing the CYCLE STOP button during automatic operation will halt all feed movements of the machine. It will not stop the spindle RPM or affect the execution of tool changes on some machines. When the CYCLE STOP button is pressed, the lamp located on the button goes on, and the lamp located on the CYCLE START button goes off. When the CYCLE STOP button is pressed, all feeds are temporarily stopped; however, the spindle rotation is not affected. This button is used when minor problems are encountered, such as coolant flow direction or when checking DISTANCE-TO-GO during setup. When the problem is remedied, press CYCLE START again to resume automatic cycle operation. It is not recommended using this button to interrupt a cut because the spindle does not stop and damage to the tool may occur. When pressed during the execution of the threading cycle, CYCLE STOP will take effect after the tap is withdrawn. If the tap breaks during the threading cycle, the only way to stop the machine is

Part 2   Operation Panel Descriptions

by pressing the RESET button on the controller or the EMERGENCY STOP button.

Prg Stop-Program Stop

When a Program Stop is commanded in the program by the program word M00, automatic operation is stopped and the LED on this button is lit. This button does not have an ON/OFF function that affects the program stop condition. It is merely an indicator lamp for when a program stop condition is active.

Operation

The keys in the Operation section of the control are used for manual operation of the machine during setup and initial start-up. Their specific functions are described below:

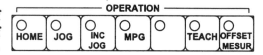

Figure 10

Home

Pressing this button on and then pressing the X or Z (X, Y or Z for machining centers), buttons causes the machine to return to the zero position for each axis in relation to the machine coordinate system.

Jog

Pressing the JOG button activates a manual feed mode that allows the selection of manual feeds along a single axes X or Z (X, Y, or Z for machining centers). With the button activated, use the Axis/Direction buttons and the Speed/Multiply buttons to move the desired axis at the chosen feed rate (in/min). On some controls Speed/Multiply is a rotary type switch that activates this function.

INC Jog

Press this button (Incremental JOG) to activate the JOG mode in Incremental steps at feed as per selection using the Speed/Multiply buttons.

MPG-Manual Pulse Generator

Pressing this button activates the manual handle feed mode for the selected axis. This handle is known as the Manual Pulse Generator.

Pressing the MPG button places the machine in the HANDLE mode. This mode enables manual control of axis movements (for X, Y or Z, or for rotational axes A, B or C) with the use of the handwheel by pressing their respective buttons. For instance, press MPG then press X and then use the Handle to move to the desired position along the X axis. By turning the handwheel clockwise, you can move the tool in a positive direction with respect to the position of the coordinate system. By turning the handwheel counterclock-

Figure 11

40

Part 2   Operation Panel Descriptions

wise, the tool is moved in a negative direction with respect to the position of the coordinate system. The handwheel contains 100 notches, each of which corresponds to an increment (distance to be moved). Turning the handwheel, you can feel the displacement from one notch to the next.

To set the magnification for the distance to be moved press one of the Speed/Multiply buttons as described below.

Caution: If the handle is rotated quickly while the magnification is set at X100 or X1K the tool will move at a rapid feed rate and a crash could occur!

Teach

This button activates the Teach-in Jog or Teach-in Handle mode. When this mode is used, the movements of the axes are recorded while in either Jog or Handle mode. Machine positions along the X, Y, and Z axes obtained by this manual operation are stored in memory as a program position and are used to create a program. These movements can then be executed as any program.

Consult the manufacturer operator's manual for detailed descriptions on its proper use. Not all controls have this feature.

Offset Mesur-Offset Measurment

When this button is pressed the OFFSET MEASURE mode is selected and the position of the tool in relation to the coordinate system is written into the tool or work offset for the active axis.

Speed/Multiply

When the INC JOG Operation mode is selected, the incremental step selected by these buttons determines the magnitude of the displacement along the chosen axis in the selected direction. When one of these buttons is pressed and released, the movement will be as follows:

X1 = a movement of .0001 inch or .0025 millimeters (mm)

X10 = a movement of .001 inch or .0254 mm

X100 = a movement of .010 inch or .254 mm

X1K = a movement of .100 inch or 2.54 mm

Figure 12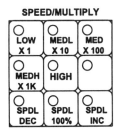

Part 2 Operation Panel Descriptions

The buttons used to select the axis and the direction of movement, are located on the upper right part of the operator panel for this controller.

For example, displace the tool along the X axis in a positive direction by the value of .010 inch, as follows.

- Press the INC JOG button.
- Press the X100 button.
- Press the button +X once.

Each time the button is pressed, a displacement of a given value results.

If the JOG mode is selected when one of these buttons are activated and the selected axis button is pressed and held in, the movement will occur at feed as indicated by LOW, MEDL, MED, MEDH or HIGH. The operator can also use the % Traverse Feed override dial to further control this feed rate.

When the MPG Operation mode is selected, the incremental step selected by the Speed/Multiply buttons determines the magnitude of the movement along the chosen axis in the selected direction.

Each button setting corresponds with the scale of the handwheel. One full revolution of the handwheel (360°) corresponds to 100 units on the scale. X means "times" the minimum increment.

When the button is pressed for;

X1, turning the wheel by one unit corresponds to the movement of .0001 inch or .0025 mm.

X10, turning the wheel by one unit corresponds to the movement of .001 inch or .0254 mm.

X100, turning the wheel by one unit corresponds to the movement of .01 inch or .254 mm.

X 1K, turning the wheel by one unit corresponds to the movement of .100 inch or 2.54 mm.

As a rule, X1 is used when you are dialing-in the zero of the workpiece and when you are determining the tool length offset.

In manual control, you must also use the AXIS/DIRECTION buttons to determine the axis of displacement.

For example, if you need to move the machine's table, with respect to the tool by 1.00 in along the X axis in a positive direction , as follows:

Press MPG Operation mode button.

Press the AXIS/DIRECTION button X+

Press the SPEED/MULTIPLY button X10.

Turn the handwheel one full revolution (100 units) and then check the value of the displacement on the screen.

Part 2   Operation Panel Descriptions

LOW
 X1

   The button, LOW indicates feed rate at a LOW speed while in the JOG Operation mode.

   The button, X1 indicates turning the handwheel by one unit cor responds to the displacement of .0001 inch or .0025 mm while in the MPG Operation mode.

MEDL
 X10

   The button, MEDL indicates feed rate at a MEDIUM LOW speed while in the JOG Operation mode

   The button, X10 indicated turning the handwheel by one unit corresponds to the displacement of .001 inch or .0254 mm.

MED
 X100

   The button, MED indicates feed rate at a MEDIUM speed while in the JOG Operation mode.

   The button, X100 indicates turning the wheel by one unit corresponds to the displacement of .01 inch or .254 mm.

MEDH
 X1K

   The button, MEDH indicates feed rate at a MEDIUM HIGH speed while in the JOG Operation mode.

   The button, X 1K indicates turning the wheel by one unit corresponds to the displacement of .100 inch or 2.54 mm.

HIGH

   This button is used to indicate the feed rate at a HIGH speed while the JOG Operation mode.

SPDL DEC-Spindle Speed Decrease

   Pressing this button causes the spindle speed to decrease.

SPDL 100%

   Spindle override 100%: Sets an override of 100% for the spindle motor speed.

SPDL INC- Spindle Speed Increase

   Pressing this button causes the spindle speed to increase.

Part 2   Operation Panel Descriptions

Spindle

These buttons are used exclusively during the manual operation of the machine for setup functions. The descriptions below explain their specific function:

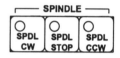

Figure 13

SPDL CW-Spindle rotation CW

By pressing this button while in one of the Operation modes; HOME, JOG, INC JOG or MPG the spindle will start rotation in the clockwise (CW) direction. The spindle RPM of the rotations are set by using the SPDL DEC, SPDL 100%, or SPDL INC buttons. The last RPM used while in this mode is retained and will restart upon pressing the SPDL 100% button.

SPDL STOP-Spindle Stop

Pressing this button stops spindle motor rotation while in one of the Operation modes listed above. Pressing this button will not stop the spindle while in any of the Execution modes.

SPDL CCW-Spindle rotation CCW

By pressing this button while in one of the Operation modes HOME, JOG, INC JOG or MPG the spindle will start rotation in the counterclockwise (CCW) direction. The spindle RPM of the rotations are set by using the SPDL DEC, SPDL 100%, or SPDL INC buttons. The last RPM used while in this mode is retained and will restart upon pressing the SPDL 100% button.

Axis/Direction

These buttons are used to select the Manual feed axis direction. Pressing these buttons executes movement along the selected axis in the selected direction by jog feed (or step feed) when the corresponding button is set to on in the jog feed mode (or step feed mode). The same is true for each of the axes where buttons are present.

-X

When pressed, this button executes movement along the X-axis in the negative direction with respect to the coordinate system.

+Z

When pressed, this button executes movement along the Z-axis in the positive direction with respect to the coordinate system.

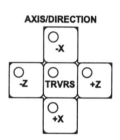

Figure 14

Part 2 Operation Panel Descriptions

-Z

When pressed, this button executes movement along the Z- axis in the minus direction with respect to the coordinate system.

+X

When pressed, this button executes movement along the X-axis in the positive direction with respect to the coordinate system.

Note: The Y axis buttons for milling machines are not depicted in the operators panel in Figure 1. The following are their descriptions.

Y-

When pressed, this button executes movement along the Y axis in the negative direction with respect to the coordinate system.

Y+

When pressed, this button executes movement along the Y axis in the positive direction with respect to the coordinate system.

TRVRS-Rapid Traverse

Use caution when using this rapid traverse function to be sure that all machine movements will not cause interference during motion!

Press this button to activate jog feed at a rapid traverse rate. Use this button to perform rapid movements along a previously chosen axis. For example, press the button TRVRS then the X positive button, the X axis will move at rapid traverse until the button is released.

Rapid Traverse Override (%)

(Not shown on this operator panel model)

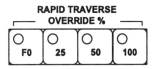

Some operator panels have a % override dial or buttons that can be used to control the rapid feed rate. This switch or these buttons reduces the rapid feed rate (G00). If it is positioned at 100, this corresponds to 100% of the feed rate that the machine can generate. Buttons are commonly in steps of 10, 25, 50 and 100%.

Figure 14b

In the above illustration F0 corresponds with 10% (as determined by parameter).

Coolant

During manual or automatic operation, these buttons may be used to activate or stop the flow of coolant.

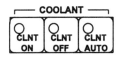

Figure 15

Part 2   Control Panel Descriptions

CLNT ON-Coolant ON

When this button is pressed the supply of coolant is started.

CLNTOFF-Coolant OFF

When this button is pressed the supply of coolant is stopped.

CLNT AUTO-Coolant Auto mode

This button is pressed to activate the automatic start and stoppage of coolant flow during program execution when called by program words M08 and M09 respectively.

Control Panel Destriptions

Control Panel

The control panel described here is quite typical of the control panels used on CNC machines. The control panel switches and buttons may be distributed differently on the panel for each individual machine, however, the purpose and function of each switch and button remains the same. Some control panels are equipped with additional buttons or switches not shown here. Definitions and applications of these buttons or switches can be found in the manufacturer instruction manuals for the machines.

The control panel is located at the front of the machine and is equipped with a CRT and with various buttons and switches, as illustrated in Figure 16.

Two items not shown on this controller that are common on many modern controls are a 3.5 floppy disk and PCMCI (Portable Computer Memory Card Interface) slot. These are both used as a file storage medium and a method for file transfer. The floppies typically hold a maximum of 1.44 mega bytes (MB) of data while PCMCI cards range

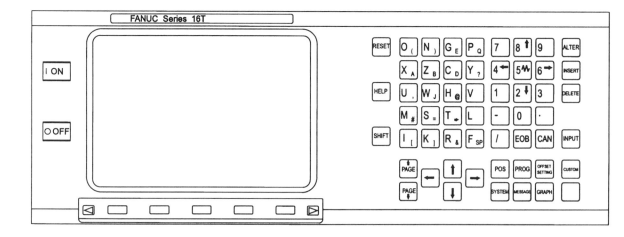

Figure 16

Part 2 Control Panel Descriptions

from 40MB to 200MB. The 1.44 MB floppy can store the equivalent of 3600 meters of paper tape. Both widely used to store part programs offset data and NC parameter information. Because of emerging technologies in computer industry, the storage and data transfer medium described here are changing and improving rapidly.

A detailed description for the use of each button and the purpose of the particular sections of the control panel are presented in the following sections:

Power-ON & Power-OFF

These buttons are used to activate/deactivate the power to the control. The ON button is typically green in color and when it is ON, the key is lit. The OFF button is typically red in color and when the power is turned OFF to the control, the key is lit. At startup of the main power to the machine, the OFF button is lit.

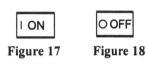

Figure 17 Figure 18

Press theses buttons to turn CNC power ON and OFF.

Notes: The control is always turned ON after the MAIN POWER switch, located on the door of the control system typically at the back or side of the machine, is ON. The control is always turned OFF before the MAIN POWER switch is turned OFF.

CRT Display

This is the television-like screen on which all the program characters and data are shown. Sizes vary from around 9 inches to approximately 15 inches. Displays are color, monochrome or Liquid Crystal Displays (LCD).

Reset Key

Pressing this button resets or cancels an alarm and can be used for cancellation of an automatic operation. An alarm can only be cancelled if its cause has been eliminated.

Figure 19

When the reset button is pressed during automatic operation, all program commanded axis feeds and spindle RPM's are cancelled.

The program will return to its starting block when this button is pressed.

Help Key

Pressing this key gives the operator a help screen s on how to operate the machine functions such as MDI key operation, or details related to an alarm that has occurred in the control.

Figure 20

Soft Keys

The soft keys have numerous functions, depending on the applications selected with other keys. The specific functions of the soft keys are displayed at the bottom of the CRT

Part 2 Control Panel Descriptions

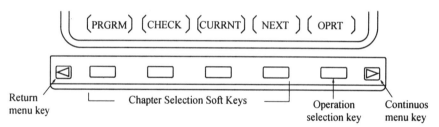

Figure 21

screen as shown in Figure 21. The purpose of the soft keys is to minimize the use of dedicated keys on the control panel.

By use of the soft keys; the machine ORIGIN counter can be reset, READ soft keys allow entering the program to memory from a punched tape or other storage medium, and the PUNCH soft key allows program readout from memory to a punched tape or other storage medium.

By pressing a soft key, the function selections that belong to it appear.

These selection choices are called chapters. These selection soft keys are the first four rectangular keys under the CRT. By pressing one of the chapter selection soft keys the screen for the selected chapter appears.

If the soft key for a target chapter is not displayed, you must press the continuous menu key located at the right end of the soft keys (sometimes referred to as next-menu). In some cases, there are additional chapters that can be selected from within a chapter.

When the desired screen is displayed, press the soft key under operation selection (OPRT) on the display for data to be manipulated.

To reverse through the chapter selection soft keys, press the return menu key located at the left end of the soft keys.

The general screen display procedure is explained above; however, the actual display procedure varies from one screen to another. For details, see the description of individual operations.

Note: The operator should consult the manufacturer manual for more specific detailed instructions on the use of the soft keys.

Address and Numeric Keys (Alpha-Numerical Keys)

This keypad of letters, numbers, and symbol characters is used to input data while writing or editing programs at the control. These keys are also used to enter numerical data and offsets into memory. Many of the keys are used in conjunction with other keys.

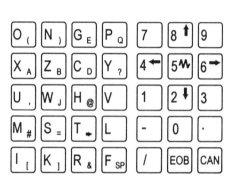

Figure 22

Part 2 Control Panel Descriptions

Shift Key

Because there is not enough space on the control for all keys necessary, some keys have two characters on them. When the letter or symbol indicated in the bottom right corner of the key is needed, the operator first presses the SHIFT key, which switches the key to that character. This sequence must be followed each time an alternate letter is needed. The shift key functions the same way as its equivalent on a computer keyboard.

Figure 23

On the display, a special character will be shown when a character indicated at the bottom right corner on the keytop can be entered.

Input Key

The INPUT key is used for MDI operation and to change the offsets. After the data are entered via the keypad the INPUT key is pressed. The data are entered into the offset register or the program for execution.

Figure 24

Cancel Key

This key is used while inputting data in the MDI mode. It is essentially a destructive backspace key and can be used to correct an erroneous entry. Press this key to delete the last character or symbol input to the key input buffer. For instance, when the key input buffer displays

Figure 25

N5X12.00Z

and then the cancel key is pressed, the address Z is canceled and

N5Xl2.00 is displayed

EOB Key-End of Block Key

This is the END OF BLOCK key and when pressed while in the MDI mode the EOB character (;) is inserted into the program at the cursor location.

Figure 26

Note: The ; symbol is never part of the program manuscript. The control system will automatically show the ; character, for every "Enter" key used on a keyboard.

Part Program Edit Keys

These keys are used to enter new program data (Insert), to make program changes (Alter), or to delete program data in memory (Delete). They are used while editing programs.

Figure 27

Part 2 Control Panel Descriptions

Function Keys

These buttons correspond to particular display modes (active mode).

By pressing any one of these buttons, the display will be switched to the corresponding screen. Then the soft keys may be used to display the needed data.

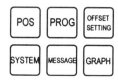

Figure 28

Press the POS key to display the Position Screen

Press the PROG key to display the program list screen

Press the OFFSET/SETTING key to display the screen used to set offsets or adjust parameter settings

Press the SYSTEM key to display the system screen

Press the MESSAGE key to display the message screen

Press the GRAPH key to display the graphics screen

CURSOR

This symbol is in the form of a blinking dash on the display, which is located below the position of a particular address while in one of the Edit modes. On many controls, the cursor highlights the whole word, for example X7.777

Example:

O0001

N1G50 X7.777 Z7.777S1000

N2T010M39

N3G96S600M03

N4 . . . .

In the above example, pressing the CURSOR button three times with the right-pointing arrow causes the cursor, located below the letter (address) G, to move to the right three spaces.

N1G50X7.777Z7.777S1000

By pressing the CURSOR button (with the arrow pointing up) repeatedly, the prompt will move to the first word of program O, which corresponds to the upper limit of cursor movement. Another fast way to return to the program head is to press the RESET key.

By pressing the CURSOR button once, with the arrow pointing down, the cursor will move down. If the cursor must be moved over a few or many words, you need not press the button repeatedly. Just press and hold this button down; the cursor automatically jumps one word at a time in the given direction. The PAGE keys allow for scrolling through long programs more effectively.

Part 2 Operations Performed at the CNC Control

Cursor Move Keys (Navigation Keys)

In order to navigate through the program, there are four keys used to move the cursor.

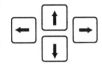

Figure 29

The right pointing arrow key is used to move the cursor to the right or in the forward direction. When this key is pressed, the cursor moves only one space each press of the button, in the forward direction.

The left pointing arrow key is used to move the cursor to the left or, in the reverse direction. Just as with the prior described case, when this key is pressed, the cursor moves only one space each press of the button, in the reverse direction.

The downward pointing arrow key is used to move the cursor downward through the program in the forward direction. When this key is pressed, the cursor moves one full line downward in the forward direction, each time.

The upward pointing arrow key is used to move the cursor upward through the program in the reverse direction. When this key is pressed, the cursor moves one full line upward in the reverse direction each time.

Page-UP/DOWN Keys

Usually the length of the program exceeds what the height of the screen will display. The CURSOR move keys can be used to scroll through the program. When you press and hold the CURSOR button with the down or up arrow, the cursor will move through the program line-by-line. A more effective method to move a large amount is to use the two PAGE keys. Using these keys will advance in the direction selected by the number of lines the screen can display. The last block of a given page becomes the first block of the next page. Use the CURSOR button with the arrow pointing up to change pages in the opposite direction.

Figure 30

Operations Performed At The CNC Control

The following explanations are for operations considered routine for operators of CNC machine tools and are given in their sequence of use.

Please note that the following procedures are specific to the type controller depicted here (Fanuc 16 or 18 series). The procedures for another type control may be similar. Be sure to consult the manuals specific to your machine tool operator and control panel.

The Machine is Turned on and Homed (Machine Zero)

Turn on the main power switch, then press the ON Power button on the controller. Most modern machine tools will automatically start-up in the HOME mode. This means that before any automatic or manual operation may begin, it will be required to Home the machine first.

If the Operation selection LED, HOME, is not lit, press it now.

Using the Axis/Direction keys, press the direction necessary to HOME the machine.

Part 2 Operations Performed at the CNC Control

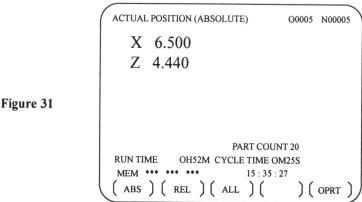

Figure 31

Many machine tools will have LED's for each axis that are lit to indicate when an axis is HOMED.

At machine start-up, a common screen displayed is ACTUAL POSITION (ABSOLUTE). If it is not displayed, press the function key labled POS then the soft key ABS. The displayed coordinate values represent the relationship between the Workpiece Zero and the Machine Zero (HOME). When the machine is HOME, press the soft key OPRT, then press ORIGIN and then press ALLEXE to zero each of the coordinate axes.

By pressing the soft keys, other display screens can be activated.

When we press the button ABS (which corresponds to position), the digital counter appears on the screen for the X and Z axes, which is the absolute coordinate system for a given workpiece (for milling machines X, Y, and Z will be displayed). The position (POS) function is assigned four display screens and can be found by pressing the soft keys labeled; ABS, REL, ALL and OPRT. The first screen corresponds to a position change in the absolute system, X, Z, as illustrated. The second screen RELATIVE (REL) corresponds to position changes in the incremental system, U and W (for milling machines X, Y, and Z). The third, ALL gives representation of all four of the displays simultaneously on one screen.

The values under MACHINE represent the distance from reference position.

Figure 32

```
ACTUAL POSITION                       O0005  N00005
        (RELATIVE)              (ABSOLUTE)
   U     6.5000            X      6.5000
   W    16.0000            Z     16.0000

        (MACHINE)          (DISTANCE TO GO)
   X     0.0000            X      0.0000
   Z     0.0000            Z    -12.0000

                      PART COUNT 20
RUN TIME      0H52M CYCLE TIME 0M25S
MEM *** *** ***        15 : 35 : 27
( ABS ) ( REL ) ( ALL ) ( HNDL ) ( OPRT )
```

Part 2 Operations Performed at the CNC Control

DISTANCE TO GO is the most significant part of the third display. The coordinates in this quarter of the display of the screen correspond to the path that will be followed by the tool in order to complete the execution of a given block of information while under automatic operation.

Example:

N20G00Z0

N22G01Z-12.000F.015

When block N22 is first read by the control, the value Z-12.000 will appear under DISTANCE TO GO in the lower right corner of the screen. After moving a distance of 1 inch, the value of coordinate Z changes to Z-11.000, and so on. The other displays, "ABSOLUTE" and "RELATIVE," correspond to the first two display screens, but this time they are smaller so that all four may fit on one screen. All of the displays may be changed to read in millimeters, with respect to machine zero, by changing a machine parameter or by using a program G-code.

Figure 33

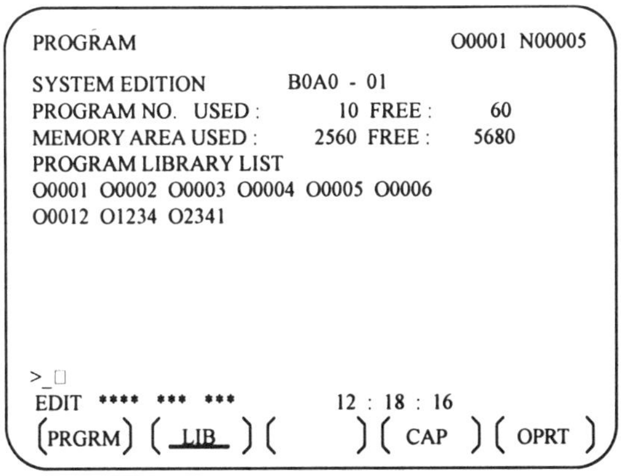

```
PROGRAM                                O0001 N00005
SYSTEM EDITION        B0A0 - 01
PROGRAM NO. USED :      10  FREE :      60
MEMORY AREA USED :    2560  FREE :    5680
PROGRAM LIBRARY LIST
O0001 O0002 O0003 O0004 O0005 O0006
O0012 O1234 O2341




>_
EDIT  ****  ***  ***            12 : 18 : 16
(PRGRM) ( LIB ) (      ) ( CAP ) ( OPRT )
```

A Program is Called From Memory

The program may be in the program directory but not activated for automatic operation. Follow these directions to activate a program.

- Press the EDIT button to enter the EDIT mode
- Press the PROG function key
- Either the program contents or program file directory will be displayed
- Press the OPRT soft key
- Press the rightmost (next-menu key) soft key
- Use the keypad to enter the desired program number proceded by the letter address O
- Press the FSRH (forward search) and the EXEC soft keys

The program will now be in the active status and ready to use for automatic operation.

Part 2 Operations Performed at the CNC Control

Program is Loaded Into Memory

Follow the directions below to load a program into the controller from an NC tape.
Be sure that the input device is ready for reading (tape entry to tape reader if used).

- Press the EDIT button on the operator's panel to enter the Edit mode
- Press the PROG function key
- Either the program contents or program file directory will be displayed
- Press the OPRT soft key
- Press the rightmost (next-menu key) soft key
- Use the keypad to enter the desired program number to load preceded by the letter address O
- Press the READ soft key and then the EXEC soft key

The program is now loaded into the controller's memory.

Program is Saved to an Offline Location

(NC Tape, Floppy, PCMCI card, or PC hard disk)
Follow the directions below to save a program to an NC tape.
Be sure that the output device is ready for output.
If the NC tape output is EIA or ISO, it needs to be specified by using a parameter.

- Press the EDIT button on the operator's panel to enter the EDIT mode
- Press PROG function key

Either the program contents or program file directory will be displayed.

- Press the OPRT soft key
- Press the rightmost (next-menu key) soft key
- Use the keypad to enter the desired program number to save preceded by the letter address O
- To save all programs stored in memory Press –9999
- To save multiple programs at one time enter their program numbers separated by a coma like this; O????, O????.
- Press the PUNCH and then EXEC soft keys

The program will be saved.

A Program is Deleted from Memory

- Enter the EDIT mode

Part 2 Operations Performed at the CNC Control

- Press the PRGRM soft key
- The program directory will be displayed
- Press the OPRT soft key
- The screen with soft keys labeled F SRH, READ, PUNCH, DELETE and OPRT will be displayed
- The program directory is displayed only while in the EDIT mode
- Press the DELETE soft key
- Enter the program file number (preceded by the letter address O) you wish to delete
- Press the EXEC soft key

The file is deleted.

To delete all programs from memory use the following directions for the last three steps:

- O-9999
- DELETE
- EXEC

MDI Operations

The operator may input small programs via the keypad at the control. The size of the program is limited to 10 lines on the control described here and is determined by the parameter setting from the manufacturer. It is an excellent method of executing simple commands like tool changes, controlling the spindle RPM and its rotation direction, etc. To enter the MDI mode of operations:

- Press the MDI button on the operator panel
- Press the PROGRAM function key
- Enter the desired program number preceded by the letter address O
- Enter the data to be executed by using the methods described in PROGRAM EDITING FUNCTIONS

As soon as the program number is entered, you can begin to enter program data. If a program number is not input, the control assumes O0000, and the data may be entered. Each block ends with end of block (EOB) character (;) so that individual blocks of information can be kept separately.

N1G50S1000;

- Press the EOB function key to insert the semicolon at the end of each line.
- Press the INSERT edit key.
- Press CYCLE START

55

Part 2 Operations Performed at the CNC Control

If a typographical error is made while entering a given block, you can eliminate the error by pressing CAN key to CANCEL the error and then entering the correct value.

The MDI program may be executed just as with automatic operation and the same control functions apply except that an M30 (tape rewind) command does not return the control to the program head instead M99 is used to perform this function. Please refer to the machine tool manufacturer manual for specific instructions.

Erasure of an entire program created in MDI mode may be accomplished as follows:

- Use the alphanumeric keypad to enter the address O
- Press the DELETE key on the MDI panel.

The same result may be accomplished by pressing the RESET key.

Also the program will be erased when execution of the last block of the program is completed by single-block operation.

To perform an individual MDI operation, use the methods described above. For the control described here the use display screen shown in Figure 34.

Example 1:

- Turn on the spindle at 500 RPM in the clock wise direction.
- Key in the following command:
- S500M03
- EOB
- INPUT
- CYCLE START

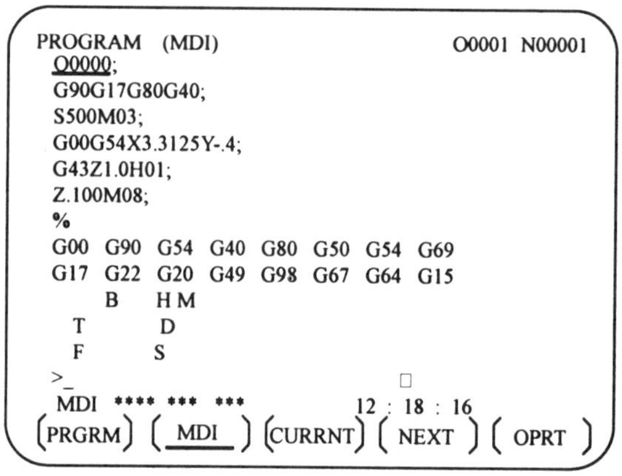

```
PROGRAM  (MDI)                    O0001 N00001
O0000;
G90G17G80G40;
S500M03;
G00G54X3.3125Y-.4;
G43Z1.0H01;
Z.100M08;
%
G00  G90  G54  G40  G80  G50  G54  G69
G17  G22  G20  G49  G98  G67  G64  G15
         B      H M
   T            D
   F            S
>_                              □
 MDI **** *** ***         12 : 18 : 16
(PRGRM) ( MDI ) (CURRNT) ( NEXT ) ( OPRT )
```

Figure 34

Example 2:

Follow these directions to position tool number 5 to the active position on the turret (or to install tool 5 into the spindle on a milling machine).

- Key in the following command: T05M06 (or T5M06 for a mill)
- EOB
- INPUT
- CYCLE START

Part 2   Operations Performed at the CNC Control

Measuring Work Offsets, Turning Center

It is necessary to establish a relationship between the machine coordinate system and the workpiece coordinate system. The following directions are given to input the measured values for the workpiece zero to the controls Work Coordinates offset page.

Measure The Z Axis Work Coordinate

Manually position the cutting tool and make a cut on the face of the workpiece.

- Without moving the Z axis, stop the spindle and move the tool away from the part in X-axis direction
- Measure the distance along the Z axis from cut surface to the desired zero point
- Press the WORK soft key to display the WORK COORDINATES display screen
- Position the cursor to the desired Workpiece offset to be set
- Use the letter address key Z to select the axis to be measured
- Use the value of the measurement taken to input the Z axis Work Coordinate
- Press the MEASUR soft key

The Work Coordinate for the Z axis will be input.

Measure the X Axis Work Coordinate

- Manually position the cutting tool and make a cut along Z axis to create a diameter on the workpiece
- Without moving the X axis, stop the spindle and move the tool away from the part in Z axis direction.
- Measure the diameter you just cut on the workpiece
- Use the value of the measurement taken to input the X axis Work Coordinate
- Follow the same procedure for setting the Z axis Work Coordinate value as stated above

The Work Coordinate for the X axis will be input.

Measuring Work Offsets, Machining Center

Following is the procedure for setting the Work Offsets for each workpiece coordinate system G54 to G59. When the values are known you can:

- Press the OFFSET/SETTING function key.
- Press the WORK soft key.

The WORK COORDINATES setting screen is displayed as shown in the Figure 35. There are two display screens needed to handle the six offsets G54 to G59. To display a desired page, follow either of these two ways:

Part 2 Operations Performed at the CNC Control

Figure 35

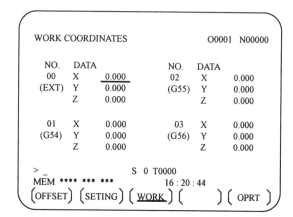

- Press the PAGE up or PAGE down keys until the desired offset shown
- Enter the offset number, G5?
- Press the NO.SRH soft key

To change the coordinate values of the offsets, use the following method:

- Use the arrow keys to position the CURSOR on the appropriate offset number.
- Use the alphanumeric keypad to enter the new value for the offset.
- Press the INPUT soft key.

Note: When the INPUT key is used to enter values, the amount entered will replace any amount in the register. When the +INPUT key is used the existing amount in the offset register will be added or subtracted, whichever applies, by the amount entered into it.

The value entered here is the zero or origin for the workpiece coordinate system.

To change an offset by a specific amount, use the alphanumeric keypad to enter the desired value then press the +INPUT soft key.

Measured Values

Work Offsets can be measured manually by positioning the edge-finding tool to contact with the workpiece zero-surface in both X and Y axes sequentially. This procedure is called Edge Finding and is nearly always the perpendicular edges (secondary and tertiary datum) of the workpiece that is referenced.

Following is a description of the Work Offset measuring procedure:

- Position the machine to HOME
- Use the procedure above to find the WORK COORDINATES setting display screen
- Use the arrow keys to position the cursor on the offset you wish to use
- Press 0 INPUT for the X value

Part 2 Operations Performed at the CNC Control

- Press 0 INPUT for the Y value
- Install an Edge-Finding tool into the spindle using MDI or manually
- Start the spindle RPM clockwise at approximately 1000 by using MDI or manually
- Manually position the tool tip edge to contact the workpiece zero-surface along the X or Y axis
- Use the alphanumeric keypad and press X or Y and then INPUT, to enter the axis to be measured

The desired axis should be blinking on the display screen and the soft key options NO. SRH and MEASUR shown.

- Press the MEASUR soft key. The absolute position value will be input to the offset.
- Manually retract the Edge-Finding tool and repeat the same operation for the remaining axis. In most cases, the operator will be required to input the difference between the value input and the Edge-Finder radius (typically 0.100 or 3mm) before automatic operation can be executed.

Figure 36

```
OFFSET                              O0001 N00000
  NO.      X         Z         R         T
  001    0.000     6.500     0.000       0
  002    0.000     0.000     0.000       0
  003    2.500    -6.000     0.000       0
  004    0.000     0.000     0.000       0
  005    0.000     0.000     0.000       0
  006    0.000     0.000     0.000       0
  007    0.000     0.000     0.000       0
  008    0.000     0.000     0.000       0
ACTUAL POSITION (RELATIVE)
     U    4.200          W    8.000
>  _
MDI  **** *** ***           16 : 20 : 44
(OFFSET) (SETING) ( WORK ) (       ) ( OPRT )
```

Tool Offsets are Measured

Turning Center Tool Offsets

On Turning Centers, the tool offsets are measured in two directions: Z and X.

These values represent the difference between the reference position (machine home) of the tool turret used in programming and the actual position of a tool tools tip used as the programmed tool point. The amount of Tool Nose Radius is input on the offset page where R is indicated for each tool. An incorrect value here will have an affect on the finished part where tapers and radii are turned.

Measured Values

If the position register commands (G50 or G92) are used, the values for each tool that have been measured will be input into the program for each tool with the G50 or G92 command.

Part 2   Operations Performed at the CNC Control

The more commonly used method today is to input these values into the offset register for each tool. Follow these steps for input of the measured tool offset value:

Figure 37

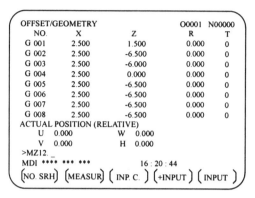

Measure the Z Axis Offset

- Manually position the cutting tool and make a cut on the face of the workpiece
- Without moving the Z axis, stop the spindle and move the tool away from the part in X axis direction
- Measure the distance along the Z axis from cut surface to the desired zero point
- Use this value to input the Z axis offset for the desired tool number, with the following procedure:
- Press the OFFSET/SETTING function key
- Press the OFFSET soft key until the required tool offset compensation display screen is found.
- Use one of the search methods or use the cursor keys to move the cursor to the offset number to be set
- Use the alphanumeric keypad to select the letter address Z
- Use the alphanumeric keypad to key in value of the measurement taken
- Press the MEASURE soft key

The difference between measured value and the coordinate will be input as the offset value.

Measure the X Axis Offset

- Manually position the cutting tool and make a cut along Z axis to create a diameter on the workpiece
- Without moving the X axis, stop the spindle and move the tool away from the part in Z axis direction
- Measure the diameter just cut on the workpiece
- Follow the same procedure for setting the X offset value as stated above

Part 2 Operations Performed at the CNC Control

Apply this method for all of the remaining tools used in the program.

The offset values are automatically calculated and set.

Tool Sensor Measuring

On some newer machines, a method of tool-offset measurement exists where a tool sensor is used as opposed to machining the diameter and face of the material.

In this case, all of the programmed tools are manually positioned to contact the sensor for each axis and the offset values are automatically input to the control. The operator still must manually enter Tool Nose Radius compensation values.

Adjusting Wear-Offsets for Turning Centers

Wear-Offsets are used to correct the dimensions of the workpiece that change because of cutting tool wear.

For a Turning Center, the X direction offset corresponds to the diameter, for example if the X wear offset for a tool is .01, an incremental change of minus .01 refers to a decrease of the diameter by .01 and an incremental change of plus .01 refers to an increase of the diameter by .01.

To adjust the WEAR-offsets:

- Press the OFFSET/SETTING button; until the screen display shown in Figure 38 appears.

Figure 38

```
OFFSET                                    O0001  N00000
NO.    GEOM (H)   WEAR (D)   GEOM (H)   WEAR (D)
001     0.000      0.000      0.000      0.000
002     0.000      0.000      0.000      0.000
003     0.000      0.000      0.000      0.000
004     0.000      0.000      0.000      0.000
005     0.000      0.000      0.000      0.000
006     0.000      0.000      0.000      0.000
007     0.000      0.000      0.000      0.000
008     0.000      0.000      0.000      0.000
ACTUAL POSIITON (RELATIVE)
    X    0.000       Y   0.000
    Z    0.000
>_
MEM ****  ***  ***              16 : 20 : 44
(OFFSET)  (SETING) (     ) (     ) ( OPRT )
```

Examples of adjusting Wear-Offsets

- For the following examples the operator should display the OFFSET screen for WEAR offsets and the cursor should be positioned to the tool and axis requiring adjustment.

Method 1: The Absolute System

If after machining the workpiece shown in Figure 39 the measured external diam-

Part 2 Operations Performed at the CNC Control

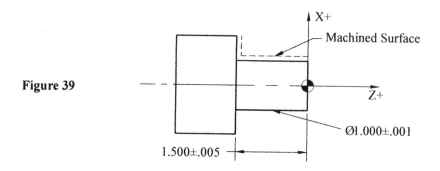

Figure 39

eter exceeds the value of tolerance (for example, 1.003), enter the offset with a negative sign assigned to the value -.003 in the wear offset.

- Press X
- Key in -.003
- Press INPUT

Then, after machining a few more pieces, the diameter increases due to tool wear. If the measured diameter is 1.002, enter the offset as follows:

- Press X
- Key in -.005
- Press INPUT

Please note that it was necessary to add a value of .002 into memory to the previously entered value of .003. A similar approach is applicable in the direction of the Z axis.

If the measured length is 1.492, then the value of the offset entered is -.008.

- Press Z
- Key in -.008
- Press INPUT

A new measured length of 1.494 gives an entered value of the offset of -.006.

- Press Z
- Key in -.014
- Press INPUT

Method 2: The Incremental System

To gain a better understanding, let us examine identical cases. The measured value is = 1.003.

Part 2   Operations Performed at the CNC Control

Offset: U

- Key in -.003
- Press INPUT

Following that, the diameter is = 1.002.

- Press U
- Key in -.002   (on the screen)
- Press INPUT (X-.005)

And Z = 1.492

Offset: W

- Key in -.008
- Press INPUT

After machining a few pieces, Z = 1.49.

Offset: W

- Press W
- Key in -.006   (on the screen)
- Press INPUT (Z-.014)

Machining Center Tool Offsets

Tool Length Offsets "TLO", are called in the program by the H word. The values input into the corresponding TLO # are needed for proper positioning of the tool along the Z axis. Similarly, the Cutter Diameter Compensation "CDC" values are entered on the offset page and are called in the program by the D word. These compensations are important for proper radial (X, Y) positioning of the tool. If the values are known, the following sequence can be used to input them into the offset page. When the setup values are known, you may:

- Press the OFFSET/SETTING function key
- Press the OFFSET soft key. It may be necessary to press this key several times until the desired offset display is present
- Use the arrow direction or page keys to position the cursor to the tool number to be set

The search method may be used also by entering the tool number whose compensation is to be changed and the pressing the NO.SRH soft key.

Enter the numerical value of the offset (including sign) and press the INPUT soft key. To add or subtract from an existing offset value key in the amount (a negative value to reduce the current value) then press the +INPUT soft key.

Diameter compensation values are input as known after measuring their actual size. Depending on the parameter setting for the specific machine used, the value is

Part 2 Operations Performed at the CNC Control

entered as either tool diameter or radius. Consult the appropriate manufacturer operation manual for exact conditions.

Measured Values

Tools length offsets can be measured by manually positioning the tool tip to contact with the workpiece zero-surface (Z axis). This procedure is called "Touching-Off" and is nearly always the topmost surface, primary datum, of the workpiece. All tools used in the program must have their offsets recorded on the offset page.

If there is not a value in the offset register for a programmed tool, the control will not execute for that tool call, an alarm will occur and the machine will stop. If a value of zero is in the offset register the control will accept the zero offset and over travel will result. Conversely, if a value in the offset register is incorrect, the control will execute the tool call as if it were correct and the result could be a collision.

For this reason, it is a good idea to delete tool offset data from the offset register when the tool for which it was intended is removed. To do this, follow the directions above to search to the tool and erase the tool data. Do not input any value for the offset register.

Following is a description of the tool offset measuring procedure:

- Manually position the tool tip to contact the workpiece zero-surface (Z axis)
- Press the POSITION function key
- Press this key several times to get to the ACTUAL POSITION (RELATIVE) display screen.
- Use the alphanumeric keypad and press Z and then INPUT, to enter the axis to be measured. The axis should be blinking on the display screen and the soft key options PRESET and ORIGIN shown
- Press the ORIGIN soft key. The value in the RELATIVE position display will be changed to 0.
- Press the OFFSET/SETTING function key several times until the offset page for tool compensation is displayed
- Manually position the tool tip to contact the workpiece zero surface (Z axis)
- Use the arrow direction keys to or search method described above to position the cursor to the desired offset
- Use the alphanumeric keypad and press Z
- Press the INPUT soft key

The relative Z value for the tool offset will be input to the offset register. Repeat for each tool used in the program.

Adjusting Wear-Offsets for Machining Centers

For milling machines, the wear offset is assigned only in the direction of the Z axis for tool length compensation. Variations in the X and Y axes are compensated for by adjusting the Cutter Diameter Compensation CDC values. The method for inputting adjustment data is similar to Adjusting Wear-Offsets for Turning Centers.

Part 2 Operations Performed at the CNC Control
Tool Path Verification of the Program

One of the optional features of modern controllers that help the operator verify that the program is ready to use is the graphic display of the programmed tool path shown in Figure 40. This visual representation of provides yet another check enabling the operator to catch any errors before machining takes place. Follow these steps to access this display screen.

- On the controller, press the GRAPH function key
- Press the GPRM soft key

The graphics parameter screen will be displayed (not shown). Use the cursor to position to each parameter and the INPUT key to insert all of the required data.

- Press the GRAPH soft key

Simulation of the programmed tool paths will be displayed on the screen.

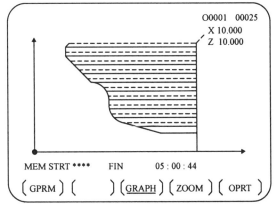

Figure 40

The operator has the added ability to adjust the magnification, change views, and display a solid model of the workpiece on the display.

Dry Run of Program

Under the DRY RUN condition the tool is moved at the rapid traverse feedrate regardless of the feedrate in the program. The actual feedrate is determined by a parameter setting and also by a rotary dial on the operation panel.

This function is used for checking the programmed movements of tools when the workpiece is not present in workholding device. The rapid traverse rate can be adjusted by using the rapid traverse feed override or by pressing F0, 25, 50, 100% buttons on the controller. Consult the manufacturer manual for specific instructions.

CAUTION!
THIS FUNCTION IS NOT INTENDED FOR METAL CUTTING.

Another form of DRY RUN is to execute the program cycle without a part mounted in the workholding device at the programmed feedrates.

Execution in Automatic Cycle Mode

Shown in Figure 41 is the program check display screen with descriptions of each content area.

When all the prior steps have been completed, the program is ready to be executed under automatic cycle. A helpful and informative display screen to use is the Program

Part 2 Operations Performed at the CNC Control

Check. This display shown in Figure 41 is convenient because the operator can see program lines as they are called, the Absolute Position display, a Distance-To-Go display and all of the active commands.

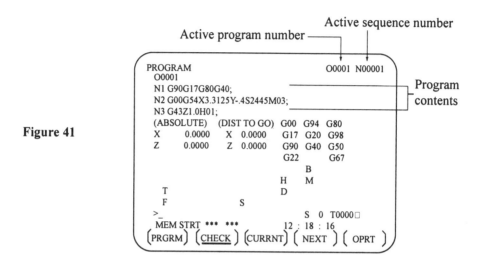

Figure 41

To begin execution in the automatic cycle mode follow these steps:

- Be sure that the desired program is in the control and active and that all set-up procedures have been completed. Use the steps above to activate if it is not.
- Press the RESET key on the controller
- Press the PRGM soft key on the controller
- Press the CHECK soft key on the controller
- Press the AUTO button on the operator panel
- Press the CYCLE START button on the operator panel

The automatic cycle will begin.

DNC Operation

Occasionally it will be necessary to run the program from a remote storage device (Floppy Disc, PCMCI card or a computer hard disk). This is typical for instances where the program is very large and will not fit in the control system memory. Direct Numerical Control (DNC) allows the program to be executed from the offline storage location. When using the computer hard drive method, the offline Personal Computer (PC) is required to have the necessary communications software and RS232 cabling hardware connected. Please consult the manufacturer manual for specific instructions. To execute a DNC operation follow these steps:

- Call the program number to be executed by one of the search methods
- Press the RMT button on the operators panel to set REMOTE execution mode

Part 2   Operations Performed at the CNC Control

- Note that this button is not shown on the controller we depict
- Press CYCLE START to begin automatic operation

Program Editing Functions

Editing of part programs includes inserting, deleting, and replacement of program words.

Understanding the techniques for program number searching, sequence number searching, word searching and address searching is required before any editing of the program can begin. The control needs to be in the proper mode and the program called.

The following are descriptions of how each is accomplished.

- Press the RESET key
- Press the EDIT key to activate the EDIT mode
- Press the PROG function key
- If the program to be edited is not active you must call it now. Follow the directions as stated under "A Program is Called From Memory"

Setting the Program to the Top

By pressing the RESET key while in the EDIT or MEMORY mode, the active program will be returned to the beginning line or program top (program head).

The second method is accomplished by doing the following:

- Press the letter address O while in either the MEMORY or EDIT modes.
- Using the alphanumeric keypad input the program number.
- Press the O SRH soft key.

For the third method:

- From the EDIT or MEMORY mode
- Press the PROG function key
- Press the OPRT soft key
- Press the REWIND soft key
- Search of the Program

Cursor Scanning

The program may be scanned to an editing location by using the cursor and the page keys. Follow the directions as stated under arrow direction, page and cursor. Using cursor scanning is not the most efficient method of searching through the program for edit locations if the program is very large.

Part 2 Operations Performed at the CNC Control

Sequence Number Searching

If the sequence number is known in the program requiring editing the operator can search directly to that location by following these steps.

- From the EDIT mode, use the alphanumeric keypad to input the sequence number preceded by the letter address N

- N????

- Press the SRH soft key forward or reverse for the direction needed

The cursor will be moved to the identified sequence number.

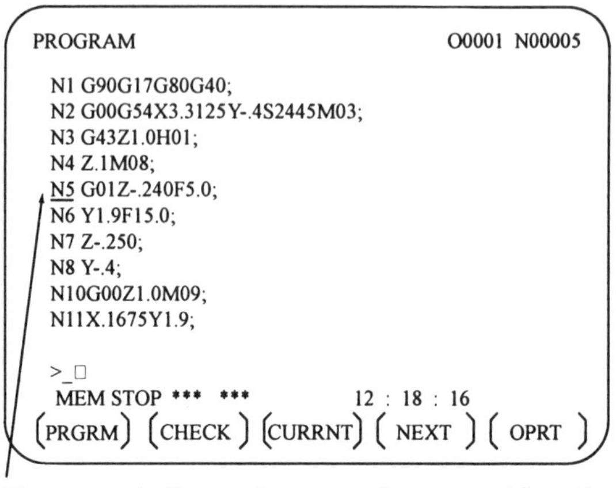

```
PROGRAM                              O0001 N00005
N1 G90G17G80G40;
N2 G00G54X3.3125Y-.4S2445M03;
N3 G43Z1.0H01;
N4 Z.1M08;
N5 G01Z-.240F5.0;
N6 Y1.9F15.0;
N7 Z-.250;
N8 Y-.4;
N10G00Z1.0M09;
N11X.1675Y1.9;

>_
MEM STOP ***  ***           12 : 18 : 16
(PRGRM) (CHECK) (CURRNT) ( NEXT ) ( OPRT )
```

The cursor indicates the currently executed location

Figure 42

Word Searching

Much like sequence number searching, the operator can search to a specific word in the program. For instance, to search to a specific word in the program, like T5 follow these directions:

- From the EDIT mode using the alphanumeric keypad key in the letter address T

- Press the number 5

- Press the SRH soft key forward or reverse for the direction needed

The cursor will be moved to the identified word T5.

Address Searching

Like word or sequence number searching, the operator can search to a specific address in the program. For instance, to search to a specific address in the program, like M06 follow these directions:

- From the EDIT mode using the alphanumeric keypad key in the letter address M

- Press the SRH soft key forward or reverse for the direction needed

The cursor will be moved to the identified address of word M06 or the first instance of the M-address that is found.

Inserting a Program Word

From the EDIT mode use a searching method to scan the program to the word immediately before the word to be inserted.

- Use the alphanumeric keypad to key the address and the data to be inserted.

- Press the INSERT Edit key.

- The new data are inserted.

Part 2   Operations Performed at the CNC Control

Example: To insert the program word Z.2 on sequence number N4 of the program listed below:

- Press the EDIT key
- Press the PRGRM soft key
- Key in the word, X1.2
- Press the SRH soft key in the forward direction
- Key in the new word, to insert, Z.2
- Press INSERT

    O1234

    N1G50S1000

    N2T0100

    N3G96S600M03

    N4G00X1.2

The result will be as follows: N4G00X1.2Z.2

Altering Program Words

From the EDIT mode, use a searching method to scan the program to the word to be altered.

- Use the alphanumeric keypad to key the new address and the new data to be inserted.
- Press the ALTER Edit key
- The new data are changed
- To change the program word, Z.2, in the example to Z.3 follow these steps
- Press the EDIT key
- Press the PRGRM soft key
- Key in the word, Z.2
- Press the SRH soft key in the forward direction
- Key in the new word, to insert, Z.3
- Press ALTER

Deleting a Program Word

From the EDIT mode use a searching method to scan the program to the word that needs to be deleted. Then press the DELETE Edit key.

To delete the program word, Z.3 from the example follow these steps.

- Press the EDIT key

Part 2   Operations Performed at the CNC Control

- Press the PRGRM soft key
- Key in the word Z.3
- Press the SRH soft key in the forward direction
- Press DELETE

SETTING

The SETTING soft key is accessible by first pressing the OFFSET SETTING function key on the control panel. The setting soft key is the second key from the left. The PAGE key may be used to view multiple display screens. By pressing this key, access is gained to the Parameter setting, sequence number comparison setting, run time and parts count display, etc. By pressing the soft key this display page allows the operator to adjust settings to enable or disable parameter writing, set automatic insertion of program sequence numbers, change from inch to metric units, and set any mirror image data needed.

From the MDI mode:

- Press the SETING soft key. The SETTING display will appear on the screen
- Set values as necessary for the desired results

Note that some of the information about the basic parameters of the machine is shown on the screen. To change any of these parameters, perform the following steps.

The mode is set to MDI.

- Press the function key OFFSET SETTING
- Press the soft key SETTING
- Use the PAGE keys to display the desired screen

The arrow keys to position the CURSOR under the parameter that you wish to change.

- Enter the desired new value
- Zero (0) or 1 is entered, where 1 indicates the ON condition and 0 the OFF condition
- Then press INPUT

PARAM

This soft key is used to access the display of a set of codes that control certain constants assumed during programming. These codes are numerical values that usually exclude a decimal point. Also, they are sometimes hexadecimal numbers representing an ON/OFF condition for each place in the number having multiple functions. Access to most parameters is not allowed without unprotecting their access. This is done through SETTING as described above. Consult the manufacturer manual for specific directions

Part 2   Operations Performed at the CNC Control

to unprotect parameters.

An example of a parameter setting is the amount a drill will retract during the chip breaking process that is assigned to canned cycle G83.

CAUTION!
Parameters should only be changed when the results of such change are understood completely. The changes will affect all programs that are executed.

To access the Parameters for adjustments, follow these steps:

- Enter the MDI mode
- Press the SYSTEM button
- Press the soft key PARAM
- Move the cursor to the desired parameter screen by using the PAGE keys
- Move the cursor to the desired position to change by using the arrow keys
- Enter the new value of the desired change
- Press the soft key INPUT

DGNOS

By pressing the DGNOS soft key the diagnostics screen is displayed. It defines a set of coded digits, which allow a quick determination of the cause of any machine damage or required maintenance. Maintenance personnel use this display screen to obtain needed information.

Tape Code

The tape code is usually made of 1-in-wide paper that includes eight channels with several combinations of punched holes. Part programs written on process sheets are punched EIA (Electronic Industries Association) or ISO (International Organization of Standardization) codes. On older machines that use this method of program file transfer, the operator must be sure the control is switched to the same code as used on the tape. With the EIA tape coding system an odd number of holes is punched, whereas with the ISO system an even number of holes is punched.

Common Operation Procedures

In this book, we want to include explanations concerning the situations that may arise during actual machining. We will concentrate on the procedures that should be fol-

Part 2   Common Operation Procedures

lowed when repetition of particular parts of the program for a specific tool is required. We will also review cases when there is a need to use an EMERGENCY STOP button and recovery from this condition.

CNC Turning Center Program

    O1234

    N1G50S2000

    N2T0100

    N3G96S400M03

    N4G00X1.25Z.2T0101M08

    N5   ...   .

    N6   ...   .

    ...   .

    N17M01   (OPTIONAL STOP)

    N18G50S1000

    N19T0200

    N20G96S200M03

    N21G00X.75Z.1T0202M08

    N22   ...   .

    ...   .

    N39M01   (OPTIONAL STOP)

    N40G50S2500

    N41T0300

    N42G96S600M03

    N43G00X2.2Z.05T0303M08

    N44   ...   .

    ...   .

    N45M30

Using the above program example, let us review a procedure which should be followed if you need to repeat a part of the program for tools T01, T02, and T03.

Part 2 Common Operation Procedures

Case No. 1 Problem:

Execution of the program was interrupted in block N17 and you need to repeat from the beginning all operations performed by tool T01.

Solution:

From the AUTO, MEMORY or EDIT mode

- Press the RESET button

This will cause a cancellation of NC control and return to the program to the beginning.

- From the AUTO or MEMORY mode
- Press CYCLE START

Case No. 2 Problem:

Execution of the program was interrupted in block N39, and you need to execute a program for tool T02.

Solution:

From the EDIT mode

- Press the RESET button
- Using alphanumeric keypad on the control panel and the search methods described earlier, search to block N18.
- From the AUTO or MEMORY mode
- Press CYCLE START

Note: In both cases, if you intend to execute the program to the end without interruption, the OPTIONAL STOP button must be turned OFF. However, if you intend to execute only part of the program corresponding to work of tool T01 or T02, the procedure is as follows.

- Press the OPTIONAL STOP button to the ON condition

After work is completed by the desired tool,

- Press the RESET key while in the AUTO, MEMORY or EDIT mode

The machine is ready for automatic cycle once again from the program head. The program will stop after reading an M01 code.

Case No. 3 Problem:

Execution of the whole program is completed but you need to repeat operations performed by tool T03.

Part 2 Common Operation Procedures

Solution:

From the EDIT mode

- Press the RESET button (if the program is completed and at its head the RESET is not required)
- Using alphanumeric keypad on the control panel and the search methods described earlier, search to block N40
- From the AUTO or MEMORY mode
- Press CYCLE START

Case No. 4 Problem:

Execution of the program is interrupted by the use of the EMERGENCY STOP button. In this case, follow the same procedure as mentioned above. However, you must HOME the machine to reset the machine coordinate system with respect to the X and Z axes.

CNC Machining Center Program

```
O2345
N1G40G80G90
N2G54G00X0.Y1.5S1520M03
N3G43Z1.0M08H01
N4 ...      ...     ...     ...     ...     ...     ...     ...     ...
......      ...     ...     ...     ...     ...     ...     ...     ...
N29G91G28Z0.
N30M01
N31T02
N32M06
N33G90G54G00X.5Y1.3S1500M03
N34G43Z1.0M08H02
N35G81G98Z-.47F6.0R.1
......      ...     ...     ...     ...     ...     ...     ...     ...
......      ...     ...     ...     ...     ...     ...     ...     ...
N38G91G28Z0.
N39M01
N40T03
```

Part 2   Common Operation Procedures

N41M06

N42G90G54G00X-4.125Y0.S2000M03

N43G43Z1.M08H03

N44...   ...   .

....

N55G91G28Z0

N56T01

N57M06

N58G28X0.Y0

N59M30

Case No. 1 Problem:

Execution of the program was interrupted in block N30, and you need to repeat operations performed by tool T01.

Solution:

From the AUTO, MEMORY or EDIT mode

- Press the RESET button
- Using alphanumeric keypad on the control panel and the search methods described earlier, search to block N3
- From the AUTO or MEMORY mode
- Press CYCLE START

Case No. 2 Problem:

During the work of tool T02, the tool was damaged, and you should change the tool and repeat all operations performed by this tool.

Solution:

- Press the CYCLE STOP (Feed Hold) button
- Press RESET to stop spindle rotations and coolant flow
- Change to one of the OPERATION MODES, JOG or MPG
- Move the axes to a clearance point from the part
- Press the HOME button
- Use the axis jog direction keys to HOME the axes

Part 2   Common Operation Procedures

- Change the tool and clear the wear offset for tool 2 and remeasure the tool length offset
- Press the EDIT key
- Using alphanumeric keypad on the control panel and the search methods described earlier, search to block N33
- From the AUTO or MEMORY mode
- Press CYCLE START

Note: In both cases, in order to repeat the work of the remaining tools, the OPTIONAL STOP button should be in the OFF condition.

If you need to repeat the work of only one tool, follow the steps listed below.
OPTIONAL STOP ON.
After machining is completed, for tool T01:

- Press EDIT
- Press the RESET
- Zero the machine with respect to X, Y, and Z axes
- AUTO OR MEMORY
- CYCLE START

After machining is completed, for tool T02 or T03:

- Press EDIT
- Press the RESET
- Zero the machine with respect to X, Y, and Z axes
- Using alphanumeric keypad on the control panel and the search methods described earlier, search to block N31 for T02 or N40 for T03
- AUTO OR MEMORY
- CYCLE START

Case No. 3 Problem:

Execution of the whole program is completed, but you need to repeat the operations performed by tool T03.

Solution:

- Press EDIT
- Using alphanumeric keypad on the control panel and the search methods described earlier, search to block N40
- AUTO or MEMORY
- CYCLE START

Part 2 Study Questions

Case No. 4

Every time that you have used the EMERGENCY STOP button, make sure to HOME the machine axes. Then follow the procedures listed above.

In this section of the book we have covered Operation. Please note that there are hundreds of situations possible and there is not enough space to cover everything here. The intent was to give a basic understanding of the Operation function for Turning and Machining Centers. For complete details on operation features specific to your machine, consult the manufacturer manuals.

Part II
Study Questions

1. The counterclockwise direction of rotation is always a negative axis movement when referring to the handle (pulse generator).

   T or F

2. Which display includes the programmed Distance-to-Go readouts?

3. When the machine is ON and the program-check screen is displayed, there is a list group of G-Codes displayed. What does this indicate?

4. Describe the difference between the Input and the +Input soft keys in the function.

5. Which button is used to activate automatic operation of a CNC program?

   a) Emergency Stop   c) Cycle Start
   b) Cycle Stop   d) Auto

6. Which display lists the CNC program?

   a) Position page   c) Program check
   b) Offset page   d) Program page

7. When the machine is turned on for the first time, it must be sent to its home position.

   T or F

Part 2 Study Questions

8. Which operation selection button allows for the execution of a single CNC command?

    a) Dry run   c) Block delete

    b) Single block   d) Optional stop

9. Which mode switch/button enables the operator to make changes to the program?

    a) Edit   c) Auto

    b) MDI   d) Jog

10. What does the acronym MDI stand for?

11. Which display screen is used to enter tool information?

12. If the Reset button is pressed during automatic operation, spindle rotations, feed and coolant will stop.

    T or F

13. During setup, the mode switch used to allow for manual movement of the machine axes is:

    a) Auto   c) Edit

    b) MDI   d) Jog

PART 3

Programming Computer Numerically Controlled Turning Centers

Miscellaneous Functions (M Functions)

M-Code	Function
M00	Program stop
M01	Optional top
M02	Program end without rewind
M03	Spindle ON clockwise (CW) rotation
M04	Spindle ON counterclockwise (CCW) rotation
M05	Spindle OFF rotation stop
M06	Tool change
M07	Mist coolant ON
M08	Flood coolant ON
M09	Coolant OFF
M10	Chuck close
M11	Chuck open
M12	Tailstock quill advance
M13	Tailstock quill retract
M17	Rotation of tool turret forward
M18	Rotation of tool turret backward
M21	Tailstock direction forward
M22	Tailstock direction backward
M23	Threading finishing with chamfering
M24	Threading finishing with right-angle
M30	Program end with rewind
M41	Spindle LOW gear range command
M42	Spindle HIGH gear range command
M71	Bar feed ON - start
M72	Bar feed OFF - stop
M73	Parts catcher advance
M74	Parts catcher retract
M98	Subroutine call
M99	Return to main program from subroutine

Chart 1

Miscellaneous Functions (M-Codes)

Miscellaneous functions are used to command various operations. Activating coolant flow (M08), and starting spindle rotation in the clockwise direction are two commonly used M-Codes. The code consists of the letter M typically followed by two digits. Normally, one block will contain only one M-Code function; however, up to three M-Codes may be in a block depending upon parameter settings. Most of the common M-codes are listed in the chart; however, many machine tool builders assign others for specific purposes relative to their equipment. Always consult the manufacturer manuals specific to the machine in use for pertinent M-codes. (See page 80 for chart.)

Program Stop (M00)

Spindle RPM, feeds, and coolant flow stop when this command is encountered. This function interrupts the automatic work cycle in order to allow the following:

1. In-process inspection and gauging.
2. Visual inspection of tool wear and other components.
3. Removal of chips.
4. Interruption of the cycle, in order to relocate the workpiece when the workpiece is being machined from both sides during one operation.

The control system identification of function M00 is accompanied by the following events:

1. Spindle revolutions are stopped.
2. Flood coolant flow is deactivated.
3. All feed movement is stopped.
4. The CYCLE START, light will still be on.

In the following block, M03 or M04 and M08 must be stated in the program again, to reactivate these functions. Functions G96, G97, S, and F are not canceled by M00.

Optional Program Stop (M01)

This function is the same as M00, with one difference: it is applied only by pressing or switching the OPTIONAL STOP button or toggle switch ON, for example, to stop the machine so that measurements can be taken, or to remove chips at the discretion of the machine operator.

Program End (M02)

Function M02 cancels the automatic work cycle, interrupts revolutions, feeds the coolant flow and cancels the control system of the NC. The CYCLE START, light goes off in some types of controls and the PROGRAM END, light comes on. The cycle is not repeated; repetition of a programmed operation is not possible. For this to occur, the M30 function is used. This command is primarily used on NC tape machines.

Spindle On (Clockwise) (M03)

This function signals the machine to activate the spindle with clockwise revolutions. M03 is cancelled by M04, M05, M00, M01, M02, and M30.

Spindle On (Counterclockwise) (M04)

To activate or cancel this function, follow instructions as for M03.

Spindle Off (M05)

This function is cancelled by M03 and M04.

Flood Coolant On (M08)

M08 activates the flood coolant flow.

Coolant Off (M09)

M09 deactivates the coolant flow.

Chuck Close and Open (M10 & M11)

M10 automatically closes the spindle chuck jaws and M11 automatically opens the spindle chuck jaws. M10 and M11 are used in certain cases when there is a special gripper for the insertion and removal of the workpiece from the spindle chuck. Such devices are used in automatic operations and where mass production is the primary focus.

Tailstock Quill Advance and Retract (M12 and M13)

During the process of turning long shafts, a tailstock is often used to support the shaft. A center drilling operation must be programmed first. Then, by applying functions M12 and M13, the workpiece can be supported with a center and the turning of the workpiece can be performed. When the operation is completed, the tailstock is returned to its original position by the M13 command. All operations are performed automatically, without interruption. If the time required for the extension of the tail spindle is noticeably long after function M12, apply a dwell function G04.

Rotation of the Tool Turret Forward & Reverse (M17 & M18)

M17 rotates the tool turret in the normal (clockwise) direction; M18 rotates the tool turret in the opposite direction (counterclockwise). These functions may only be used for some types of machines. Function M17 is valid at machine start-up. Function M18 implies change in direction of the rotation opposite to the one set previously.

Tailstock Direction Forward & Reverse (M21 and M22)

Programmable shifting of the tailstock as a whole is also done in some types of machines, especially if the extended length of the tailstock spindle is not sufficient to perform the operation and/or the lathe is large and has a long Z axis stroke. This is a factory option only.

Thread Finishing (M23 and M24)

M23 should be applied to the normal threading cycle when the threading tool usually exits at an angle. M24 is necessary when the ending thread is followed by a greater diameter or a recess groove. M23 is the default state after machine power is turned on.

End of Program (M30)

This function is similar to function M02. However, it also activates the rewinding of the punched tape if the program is run from punched tape (NC), or returns to the beginning of the program, if CNC is used.

Preparatory Functions (G-Codes)

Preparatory functions are programmed with an address G, normally followed by two digits, to establish the mode of operation in which the tool moves. In the following chart and throughout this text, the G-Codes listed and explained refer to the most commonly used Fanuc system, Type A, and as used on the 16T control. There are some variations in the use of the other two types, B and C, but most of the codes are identical. Consult the programming manual for the specific control prior to selection of the type when system programming. (See following page for chart.)

Part 3 Programming CNC Turning Centers

Preparatory Functions (G-Codes) specific to Turning Centers

Code	Group	Function
*G00	01	Rapid traverse positioning
G01	01	Linear interpolation
G02	01	Circular and helical interpolation CW (clockwise)
G03	01	Circular and helical interpolation CCW (counterclockwise)
G04	00	Dwell
G09	00	Exact stop
G10	00	Data setting
G20	06	Input in inches
G21	06	Input in millimeters
*G22	09	Stored stroke limit ON
G23	09	Stored stroke limit OFF
G25	08	Spindle speed fluctuation detection ON
G26	08	Spindle speed fluctuation detection OFF
G27	00	Reference point return check
G28	00	Reference point return
G29	00	Return from reference point
G30	00	Return to second, third, and fourth reference point
G32	01	Thread cutting
*G40	07	Tool nose radius compensation cancel
G41	07	Tool nose radius compensation left side
G42	07	Tool nose radius compensation right side
G50	00	Coordinate system setting/ Maximum spindle speed setting
G52	00	Local coordinate system setting
G53	00	Machine coordinate system setting
G54	14	Work coordinate system one selection
G68	04	Mirror image for double turrets ON
*G69	04	Mirror image for double turrets OFF
G70	00	Finishing cycle
G71	00	Stock removal in turning
G72	00	Stock removal in facing
G73	00	Pattern repeating
G74	00	Peck drilling cycle
G75	00	Groove cutting cycle
G76	00	Multiple thread cutting cycle
*G80	10	Canned drilling cycle cancellation
G83	10	Face drilling cycle
G84	10	Face tapping cycle
G86	10	Face boring cycle
G90	01	Outer/Inner diameter turning cycle
G92	01	Thread cutting cycle
G94	01	Face cutting cycle
G96	02	Constant surface speed control
*G97	02	Constant surface speed control cancellation
G98	05	Feed per minute
*G99	05	Feed per revolution

Chart 2

Notes: 1. In the table, G-Codes marked with an asterisk (*) are active upon startup of the machine. 2. At machine startup or after pressing reset, the inch (G20) or metric (G21) measuring

Part 3 Programming CNC Turning Centers

system last active remains in effect. 3.G-Codes of group 00 represent "one shot" G codes, and they are effective only to the designated blocks. 4. Modal G-Codes remain in effect until they are replaced by another command from the same group. 5. If modal G-Codes from the same group are specified in the same block, the last one listed is in effect. 6. Modal G-Codes of different groups can be specified in the same block. 7. If a G-Code from group 01 is specified within a canned drilling cycle block, the cycle will be cancelled just as if a G80 canned cycle cancellation were called.

Tool Function

Tool pockets, in turrets on CNC Turning Centers, are assigned coded numbers. The coding system is fixed and cannot be erased even if the machine is turned off. In the process of programming, tool function is commanded by the four digits that follow the letter address T. The first two digits are related to the tool number and its corresponding geometry offset. The remaining two digits signify the tools wear offset number, as illustrated below, for example T0101:

Tool pocket number	Tool offset number
Geometry Offset	Wear Offset
T 01	01

Tool numbers may vary from 1 to the maximum number of pockets in the tool turret; for example: from 1 to 8, etc. If there are two tool turrets, then the numbers assigned to turret No. 2 are sequenced consecutively from the No. 1 turret. For example: if turret No.1 has 8 tools, then the first pocket in turret No. 2 would be 9. The number of the tool and wear offset available may vary, depending on the specific control, numbers 1 through 32 are common. This means that each tool number may be assigned 1 out of the 32 offset numbers. If tool offset number 01 is assigned to tool 01, this rules out the possibility of assigning the same tool offset number to another tool. Conversely, the same is true for wear offsets.

Example:
Recommended	**Not recommended**
T0101	T0501
T0202	T0602

Usually, tool offset 01 is assigned to tool number 01. This convention simplifies operating procedures for the operator.

Practical Application of Tool Wear Offset

Tool wear compensation is a procedure aimed at the correction of dimensional variations along the Z or X axes caused by tool wear or deflection.

Part 3 Programming CNC Turning Centers

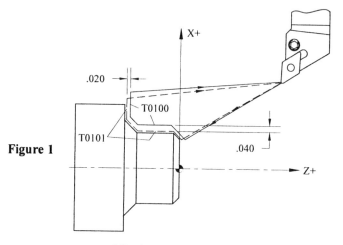

Offset is 0.040 on X axis and 0.020 on Z axis

Please note that in the above and all remaining figures where a tool is depicted in relationship to the workpiece, it will appear above the part, as is common in practice on rear turret, slant-bed style CNC Turning Centers. In reality, the tool will be a right hand style and will be facing downward. The graphical representation here shows the insert facing upward for clarity.

Keep in mind, tool geometry offset numbers refer to certain values of X and Z, and, their relationship to the workpiece zero and machine home. These geometry offsets should not be used to make adjustments for wear. Wear offsets, on the other hand, are used solely for this purpose. Small, correctional values are input to a particular tool wear offset number for X or Z to compensate for any variations. In the figure above, the solid line refers to the cutting tool path, with the X and Z values for tool wear offset as follows:

$$X = 0; Z = 0$$

The dashed line refers to the cutting tools path, with X and Z values for tool wear offset as follows:

$$X = -0.0800; Z = -0.02000$$

Notice the value of the tool offset is twice that of the dimension shown in the preceding figure, because X axis dimensions in CNC programming refer to particular diameters of the workpiece. In most cases, tool function, along with tool offset, appears in the initial phase of the program for each individual tool. The figure demonstrates that if tool function T0101 is used, the programmed tool path is changed by a displacement equal to the value of input to the wear offset.

Command T0100 initiates the cancellation of the tool wear offset. If the T0100 command is omitted, canceling the tool offset, the tool will return to the starting point with a displacement equivalent to the value of the tool wear offset. In this case, the displacement is equal to X = -0.0800 and Z = -0.02000 and is indicated on the drawing by the dashed line. Such an action will cause the tool path of each subsequent tool to be altered by the offset amount equal to the above-mentioned value. This type of programming error increases the chances for over-cutting of workpieces. Changing the tool wear offset number does not require cancellation of the previous tool offset number.

Part 3 Programming CNC Turning Centers

Feed Function

Feed function determines the amount of feed rate of the cutting tools in the machining process. Feed is programmed using the letter address "F", followed by up to four digits in the metric system and five digits in the inch system. These digits represent certain values of feed. The following examples are two methods of designating feed rate:

1. Feed rate in inches per revolutions (IPR or mm/rev) of the spindle (G99). In this case, in order to obtain feed with a certain assigned value of speed, the spindle as well as workpiece must be rotating.

Examples:

    F1.1205 in/rev.
    F0.05   in/rev.
    F0.001 in/rev.

When the spindle speed is changed after the constant surface speed per revolution has been called, the feed rate will change for a certain period of time. Therefore, feed is directly coupled with spindle speed. The following notes apply to the feed function, G99 ... F.

a. The values entered into the program for feed remain active until replaced by another feed rate, or cancelled by the G00 rapid traverse call.

b. The input value of speed is equivalent to the actual speed if the feed rate over ride on the control panel is set to 100%. See Part II, Operation, for a detailed description on Feed Rate Override.

2. Feed rate per time is measured (programmed), in inches per minute (IPM or mm/min) (G98). If feed function in the program contains feed rate per time period, then any change in the spindle speed has no effect on the feed rate because the feed rate and the spindle speed are not coupled.

Examples:

    F121.15 in/min
    F1.05   in/min
    F0.5   in/min

Generally, all CNC lathes are set to have feed per revolution of the spindle at start-up. In order to determine feed rate per minute (IPM or mm/min), the G98 function must be used, which remains effective until cancelled by function G99, or until the machine is turned off. The following notes apply to feed function, G98 ... F:

a. The values entered into the program for feed remain active until replace by another feed rate, or cancelled by the G00 rapid traverse call.

b. The input value of speed is equivalent to the actual speed if the feed rate over ride on the control panel is set to 100%. See Part II, Operation, for a detailed description on Feed Rate Override.

c. Feed functions containing feed in inches per minute, are not applicable to threading cycles.

Part 3 Programming CNC Turning Centers

Following are examples of feed functions:

F0.02	in/rev.
F0.004	in/rev.
F0.035	in/rev.
G98	
F2.0	in/min
F0.5	in/min
G99	
F0.012	in/rev.
F0.008	in/rev.

Spindle Function

Spindle function is commanded by the letter address S, followed by a number, up to four-digits, as shown below. The spindle rotation direction (M03 or M04) is typically in the same block.

S50, S150, S3000

There are two available functions applicable to the control of spindle speed. These are:

G97: Constant spindle speed
G96: Constant cutting speed

Both functions appear together with function S, for example:

G97   S500: 500 = revolutions per minute (RPM)
G96   S400: 400 = Cutting Speed (V) in ft/min (or m/min)

Constant spindle speed is applied in the case of threading cycles and in the machining of a workpiece, with the diameter remaining constant. It is also used for all operations on the centerline, like drilling, etc. If the situation calls for several changes of spindle speed in a given program, new values for the S function are assigned.

Example:
G97S1000 is active for diameter one and sets the constant spindle speed.
S800 changes the RPM for diameter two.
S300 changes the RPM for diameter three.

Constant Cutting Speed

To further examine this concept, study the following diagram of peripheral speed distribution that appears during a facing cut, using function G97 S1000. The formulas below calculate peripheral speed (cutting speed) for each diameter of 1.00, 2.00 and 3.00 inches. The result is not desirable, because of a decrease in cutting speed as the diameter gets smaller.

Part 3 Programming CNC Turning Centers

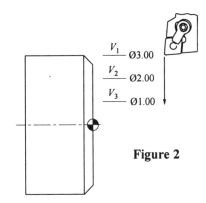

Figure 2

$$V_1 = \frac{\pi \times D \times n}{12} = \frac{3.14 \times 4.0 \times 1000}{12} = 785 \text{ (FPM)}$$

$$V_2 = \frac{\pi \times D \times n}{12} = \frac{3.14 \times 2.0 \times 1000}{12} = 523 \text{ (FPM)}$$

$$V_3 = \frac{\pi \times D \times n}{12} = \frac{3.14 \times 1.0 \times 1000}{12} = 261 \text{(FPM)}$$

where

n = RPM or rev/min

D = diameter

V = Cutting speed (FPM)

As the diagram shows, cutting speed decreases if G97 is used, as a diameter decreases and reaches zero at the centerline of the part. The question is, however, whether this phenomenon is of any advantage to us in the process of facing. Before this question is answered, look at the advice of cutting tool manufacturers who recommend specific cutting speeds for different types of machined materials. In the case of function G97, such a condition will be fulfilled only with respect to one diameter. As mentioned previously, constant cutting speed (G96) is one of the factors that can be included in programs for CNC Turning Centers.

If the cutting speed must remain constant, then the spindle speed has to increase with a decreasing diameter. Spindle speed for each consecutive diameter is calculated by the control, according to the formula $n = (12 \times V)/(\pi \cdot D)$. A closer look at this formula leads to the conclusion that, theoretically, as the diameter decreases to zero, the spindle speed increases to infinity. In reality, the spindle speed range is limited by the maximum RPM capacity of the machine.

In practical terms then, function G96 is very useful during facing and also in all cutting which involves a change in the diameter of the workpiece. As the diameter of the machined workpiece decreases, spindle speed increases, and, conversely, as the diameter increases, the spindle speed decreases.

Maximum Spindle Speed Setting (G50)

G50 S ...

When function S, is given within a program block preceded with function G50, it

Part 3 Program Coodinate Systems

refers to the maximal spindle speed that can be applied in the current operation for a given tool. As mentioned above, the spindle speed is calculated according to technological metal-cutting conditions for a given tool, or for any particular material. In some cases, the workpiece holding arrangement may require special equipment, which is mounted onto the conventional holding equipment. Such workpiece holding equipment creates conditions that do not permit utilization of the full range of spindle speeds, especially, maximum spindle speed for a given machine. Because of this fact, a maximal spindle speed for a particular operation is assigned by using the function G50. This means that, if in the machining process, metal-cutting conditions arise that require a higher spindle speed, an increase of the spindle speed will not take place. This is called "clamping the spindle speed" at a safe maximum.

Programming of CNC Lathes in Absolute and Incremental Systems

Programming in Absolute Systems

In an absolute system, input coordinates of programmed points always refer to a fixed zero coordinate point. The actual coordinates of traverse for the tool tip from point 0 to 10 are shown on the drawing in Figure 3.

In order to simplify programming as well as program readout, values of X are equivalent to each particular diameter of the workpiece.

0. X0.0 Z0.0

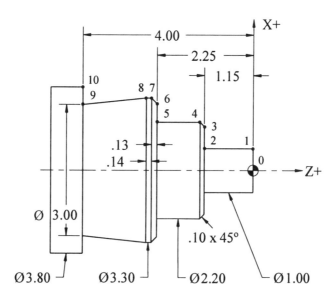

Figure 3

Part 3 Program Coodinate Systems

1. X1.0 Z0.0
2. X1.0 Z-1.15
3. X2.0 Z-1.15
4. X2.2 Z-1.25
5. X2.2 Z-2.25
6. X3.04 Z-2.25
7. X3.3 Z-2.38
8. X3.3 Z-2.52
9. X3.0 Z-4.0
10. X3.8 Z-4.0

Programming in Incremental Systems

When using an incremental system, the path of the tool from one position to the next is given in the direction of each axis. Using symbols U and W (not G90/G91 as in milling), the point displacements may be input in the direction of the X and Z axes respectively. The sign in front of the value determines the direction of movement. Values of U refer to motion in the X axis direction and refer to changes in the diameter of the workpiece. This next example illustrates the tool movement as programmed in the incremental system (refer to the same drawing).

0. U0.0 W 0.0
1. U1.0 W0.0
2. U0.0 W-1.15
3. U1.0 W0.0
4. U0.2 W-0.1
5. U0.0 W-1.0
6. U0.84 W0.0
7. U0.26 W-0.13
8. U0.0 W-0.14
9. U-0.3 W-1.48
10. U0.8 W0.0

In order to simplify the machining process, combinations of both systems (absolute and incremental) can be used. The CNC control registers the position of the tool, regardless of whichever method of programming is being used.

0. X0.0 Z0.0

Part 3 Program Coodinate Systems

1. X1.0
2. Z-1.15
3. X2.0
4. X2.2 W-0.1
5. Z-2.25
6. X3.04
7. X3.3 W-0.13
8. W-0.14
9. X3.0 Z-4.0
10. X3.8

If the value of one of the coordinates remains the same, then input only the value of the next consecutive changing coordinate.

Setting Absolute Zero of the Coordinate System (G50)

So far, the meaning of the basic terms machine zero and workpiece zero have been explained. Setting absolute zero of the coordinate system in the program is done by employing function G50 and, at the same time, inputting the value of the dimension referenced: the distance between workpiece zero and the tools cutting tip in the direction of X and Z. It must be done this way because data that refer to the X axis in CNC define the particular diameters. Please refer to Part II of this text for a detailed description on the process of setting G50 coordinates.

G50 X ... Z ...

Example:

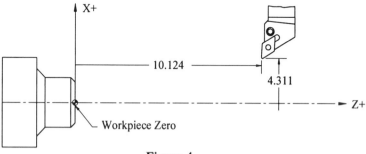

Figure 4

Notes:

1. Values of coordinates assigned to function G50 must refer to the X and Z coordinates from the workpiece zero to the tool tip.

2. Assigning smaller values for X and Z will cause an error; therefore, the tool will not approach the workpiece correctly.

3. Assigning larger values for X and Z will cause an error, possibly causing the tool

Part 3 Program Coodinate Systems

to plunge into the workpiece or the holding equipment, resulting in considerable damage to the part or machine.

4. Remember that in the case of a need for small adjustments to the X or Z axes coordinates, the Wear Offsets should be used to change the values.

5. This method is used on older controls. A more modern method incorporates the use of geometry offsets for each tool, thus, eliminating the need for the position values of each axis in the program.

The drawing below shows the case of smaller values of X and Z than required.

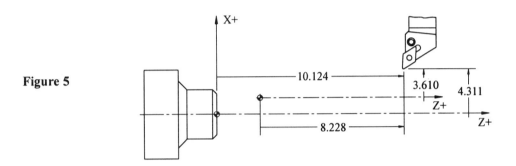

Figure 5

G50 X4.311 Z10.124   Correct

G50 X3.610 Z8.228   Incorrect

The following drawing describes the results of increasing the values of X and Z in function G50.

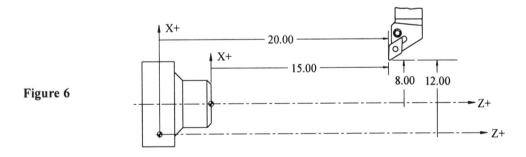

Figure 6

G50 X4.311 Z10.124   Correct

G50 X5.42 Z12.264   Incorrect

The following are examples for different shaped tools to illustrate the application of function G50.

G50 X10.8 Z8.5

G50 X14.0 Z17.0

G50 X18.0 Z21.0

Part 3 Program Coordinate Systems

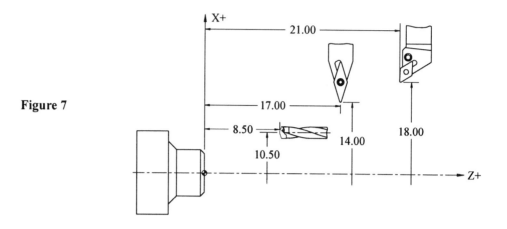

Figure 7

Review of Work Offset (G50) Measurement Technique

The value of the X coordinate assigned to function G50 refers to a diameter on which the tool tip is located with respect to workpiece zero. This assumes, of course, that the axis of the spindle is the axis of symmetry, or centerline. In order to correctly obtain this value, apply the following procedures.

1. Set the tool at machine home, or some other starting point.
2. Reset the readout displaying the value of displacement in the direction of the X axis to zero.
3. Turn on the spindle at the required speed, depending upon material type and tool type, then manually position the cutting tool to make a cut along Z axis to create a diameter on the workpiece.

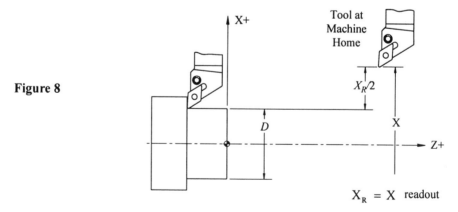

Figure 8

$X_R = X$ readout

Without moving the X axis measure the diameter of the cylindrical workpiece and add the value of this measurement to the value shown on the position readout. The result is a calculation of the desired dimension. Please refer to Part II Operation, Measuring Work Offsets, for the detailed procedure.

$$X_R = X \text{ readout}$$
$$X_R = \text{value of the displacement registered in the readout}$$

Part 3 Program Coordinate Systems

$$X = 2 \times \frac{X_R}{2} + D = X_R + D$$

For $X_R = 5.6263$ and $D = 4.383$, then G50 × X10.0093. If the diameter of the workpiece is greater than the diameter on which the tool is positioned, proceed as follows:

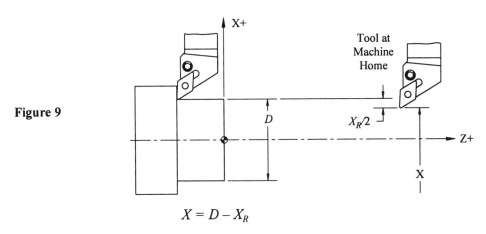

Figure 9

$$X = D - X_R$$

For $D = 20.126$ and $X_R = 3.822$, then G50 X16.304.

Attention:

The diameter obtained through initial machining used to set the values of the X coordinate assigned to function G50 is arbitrary. Measurement of the value of Z is relatively simple and is obtained from the position readout. The readout provides us with the distance traveled by the tool, along the Z axis, from machine zero to workpiece zero. Workpiece zero is chosen by the programmer in the direction of the Z axis and is typically the finished front face of the part.

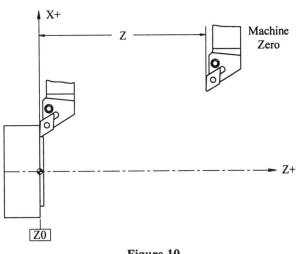

Figure 10

Values of X and Z assigned in the program to function G50 are valid only for a particular tool. They are replaced with the same function to which different values of X and Z coordinates are assigned for consecutive tools. In other words, there are new coordinate values for G50 for each tool.

The choice of sign (+ or -) for the G50 coordinates depends on the position of the tool tip with respect to the chosen workpiece zero, as shown in the following figure:

Part 3 Program Coodinate Systems

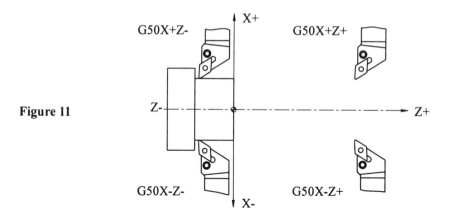

Figure 11

When creating a program, the programmer will not know the values of X and Z that are assigned to function G50 because they are measured and dependent upon the tool used. These values are variable and are input as described earlier by the operator, or setup person. Therefore, whenever the need to use these values arises in the remainder of this book, logical values will be assigned to the program. Also note that when G50S, maximum spindle speed setting is used it is independent, but can be combined into the same block.

G50 X4.311 Z10.124 S1000

In reality, this (G50) method is seldom used anymore. Because of the potential problems of incorrect entry of dimensional data into the actual program the method is prone to mistakes. The more modern technique uses the geometry offset register to store the dimensional data for each tool.

Program Creation

In order to gain a better understanding of programming, the following sample program, with explanations for every part of the program is given.

Sample Program

O1234   Program Number

N001G50X4.311Z10.124S1000

N002T0100M41

N003G96S500M03

N004G00X1.2Z.2T0101M08   Program for tool No. 1

. . .

. . .

. . .

Part 3 Program Creation

...

N016G00X20.126Z10.182T0100M09

N018M01

N019G50X18.624Z6.146S800

N020T0200M41

N021G96S600M03

N022G00X.8Z.1T0202M08   Program for tool No. 2

...

...

...

...

N030G00X18.624Z6.146T0200

N031M30

%

Program Number

The program number typically consists of a four-digit integer following the letter O. This number is used to identify the programming procedures. The range of numbers that can be used as program numbers is 0001 to 9999. Be careful not to mistake the letter O, which precedes the actual program number, for a zero.

Example of program numbers:
O0001
O0002
O0004
O1261

Block Composition

A block is basically one line of the part program. It contains the commands used to simultaneously execute operations on CNC machines. The block is composed of program words and always ends with a semicolon character known in CNC programming as the end of block (EOB). The semicolon is never part of the written or disk copy of the program. Refer to Part I, for additional information on the block format.

Examples of program blocks:
N001G50X20.126Z10.182S1000
N020T0200M41
N031M30

Part 3 Program Creation

Arrangement of the words in a given block is random, but N must be first.

Block Number

The block number (also called the sequence number) is defined by the letter N, followed by one to four or five digits, and is limited to these four or five digits. A block number provides easier access to information contained in the program. The arrangement of block numbers in a given program can be random, but typically is sequenced in increments of some amount of other than one. The most common step increments are five or ten.

Examples of block or sequence numbers:

N001

N4

N1000

N5

N10

N15

N20

Block or sequence numbers can be omitted in a block, or in the program itself. This saves storage space in the controller memory. It is important to note however, that a program searching technique requires a sequence number for the restart of a program from somewhere other than its head. The logical location then, for sequence numbers, is at tool changes enabling the restart of that tool. It is possible however, to search for a specific program word, such as T02, as well.

Part Program

The *part program* is that section of the program that contains essential information needed to control the cutting tool and auxiliary equipment.

Subprogram

The subprogram is a subordinate program. It is registered in the controller memory with the letter O, followed by a four or five-digit number, just as the main program is. In the main program, a subprogram is called by using the M98 P.... function and then M99 (called for in the subprogram), is the function that ends a subprogram.

Example subprogram:

O1234

N5	...	...	...	...
...	...	...	...	...
...	...	...	...	...

N20   M99

Part 3 Program Creation

The subprogram is called for with function M98 in the main program, while the number of the subprogram is called for with the letter P.

Example of a subprogram call:

M98P1300

Address L indicates how often the subprogram is to be repeated. The amount of repeats for L can = from 1 to 999. If the address L is omitted from the program, the subprogram is executed only once. Examine the following example to obtain a better understanding of the structure.

Example of main and subprogram structure:

Main Program	Subprogram	Subprogram
O1200	O1300	O1400
N001G50X10.0Z6.0	N001G00X2.0	N001G01Z.2
N002.......	N002......	N002 ......
N003.......	N003......	N003 ......
N004M98P1300	N004M98P1400	N004.......
N005.......	N005......	N005M99
N006.......	N006M99	
N007.......		
N008M30		

Consecutive subroutines may be called from a current subprogram by applying the above-mentioned methods. The number of subprograms linked together to form a program may be as high as 9999. In the program block N004 of the above example, execution of the program begins with subprograms O1300 and O1400. After the execution of each of these subprograms is completed, further execution of the program begins with the next consecutive block in the previous program. This is the block (N005) following the block that called the subprogram.

Program Example

Example of Program 1

For this example, only a finish cutting pass is programmed along the part geometry. The tool nose radius (for the purpose of simplification only) in this example is programmed as if to be zero.

O0001

N001G50X18.0Z10.0S1800

Part 3 Program Creation

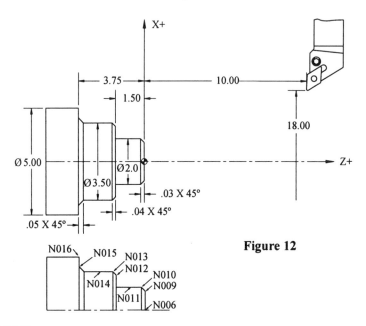

Figure 12

N002T0100M42

N003G96S600M03

N004G00X2.1Z.1T0101M08

N005G01Z0.0F.02

N006X-.03F.008

N007G00Z.1

N008X1.94

N009G01Z0.0F.02

N010X2.0Z-.03F.004

N011Z-1.5F.008

N012X3.42

N013X3.5W-.04F.004

N014Z-3.7F.008

N015U.1Z-3.75F.004

N016X5.1F.008

N017G00Z.1M09

N018X18.0Z10.0T0100M05

N019M30

Part 3 Program Creation

The following explanations are given for the individual parts of the program above.

O0001 is the program number.

N001 through N019 are sequence or block numbers.

N001G50X18.0Z10.0S1800

Block N001 contains very important information about workpiece zero. This information must be provided by the programmer or a setup person and entered at the beginning of the program with function G50. The input value X = 18.0 refers to the diameter the tool tip is set at, in the machine home position. Input value Z = 10.0 refers to the distance between the position of the tool tip and the workpiece zero position of the coordinate system on the Z axis, selected by the programmer. In this case, it is the face of the workpiece. S1800 refers to the maximum spindle speed (because it is specified within the G50 block) applicable during machining for tool No. 1(T0100).

N002T0100M42

Tool function T0100 means that tool No. 1 is in the work position. The number 00 cancels any wear offset so that no offset compensation is used for this tool, at this time. Miscellaneous function M42 provides the machine with the information that the highest spindle speed range is applicable to this tool.

N003G96S600M03

Function G96 states that the machining process will take place with a constant cutting speed of 600 ft/min, and the spindle speed will be adjusted automatically, based on the diameter of the workpiece up to a maximum of 1800 RPM. M03 refers to the direction of spindle rotation, which in this case, is clockwise. Block N003 also activates the rotation of the spindle.

N004G00X2.1Z.1T0101M08

The information contained in this block activates the execution of many of the commands related to tool motion, the flood coolant system, etc. First, the tool turret and carrier advances to a position specified by the coordinates X2.1 and Z.1 at rapid traverse, while simultaneously, the flood coolant system is activated. In addition, an applicable wear offset (coded by 01) will affect the actual path taken by the tool.

N005G01Z0.0F.02

Function G01 in the above block refers to the linear interpolation with respect to the Z axis, with the ending coordinate of Z = 0.0 and a value of feed .02 in/rev.

N006X-.03F.008

N007G00Z.1

N008X1.94

Part 3 Preparatory Functions

N009G01Z0.0F.02

N010X2.0Z-.03F.004

N011Z-1.5F.008;

N012X3.42;

N013X3.5W-0.04F.004

N014Z-3.7F.008;

N015U.1Z-3.75F.004

N016X5.1

The information found in blocks N006 to N016 refers to a change of position in the programmed points of the workpiece coordinate system which machine the part profile.

N017G00Z.109

In this block, the tool will advance to Z.1 and the flood coolant system will be turned off.

N018X18.0Z10.0T0100M05

Block N018 is also very important in the program. It commands the tool to return to its starting point identified by the G50 in block N001, cancel the wear offset and stop the rotation of the spindle.

N019M30

M30 ends the program and resets the program to its head.

Preparatory Functions(G Functions)

Rapid Traverse Function (G00)

G00 is the *rapid traverse* function; it is used for position changes without machining. G00 can be interconnected with the M, S, or T functions. G00 is a modal G-Code and will remain in effect until replaced by another command from the same group. For this example, only a finish cutting pass is programmed along the part geometry. The tool nose radius (for the purpose of simplification only) in this example is programmed as if to be zero.

Example:

O2345

N001G50X15.Z10.S2000

N002T0100M41

Part 3 Preparatory Functions

```
N003G96S500M03
N004G00X1.1Z.2T0101
N005G01Z0F.02
N006X0F.01
N007G00X.9Z.03
N008G01Z0.F.02
N009X1.W-.05F.01
N010Z-.75
N11X1.25
N012G00X15.Z10.0T0100M05
N013M30
```

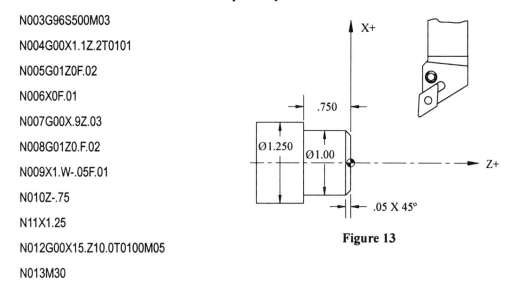

Figure 13

Caution:

If G00 rapid positioning is programmed in the direction of both axes, note that the tool will not advance to a specified point following the shortest possible path. The tool path is determined by the speed of rapid traverse with respect to each axis. In most cases, the speed for the X axis is much greater than that for the Z axis. The axis that must travel the least distance will be reached first.

The following figure 14 is a comparison between rapid and feed traverse movements when the axes are commanded simultaneously. The rapid traverse moves are indicated by dashed lines and feed traverse by solid lines.

N004 = a rapid traverse move toward the material (workpiece)

N007 = a rapid traverse move to the starting point of the chamfer

N009 = a linear feed traverse move (G01) to create the .05 X 45° chamfer

N012 = a rapid traverse move to return to the reference position identified by G50 in line N001

Workpiece holding devices (chuck jaws) quite often extend beyond the workpiece holding equipment (chuck). Therefore, careful consideration should be given to the position of work holding devices (including tailstock centers and quill extension position), so that rapid traverse paths of the tool do not interfere with them, cause a crash that may damage the machine, clamping device, or part.

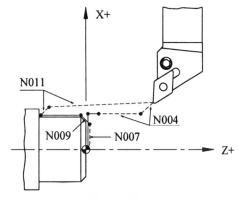

Figure 14

Part 3 Preparatory Functions

Linear Interpolation (G01)

Linear interpolation is programmed by using the G01 function and may be applied simultaneously for both axes. The G01 function commands the movement of the tool from a given position to the position with the assigned coordinates having feed rate specified by the F-Word. The block format for linear interpolation is given as follows.

G01X(U)...Z(W)...F...

The interpolator in the control system calculates various speeds for the motion axis so that the resulting speed is equivalent to the programmed feed rate. For this example, only a finish cutting pass is programmed along the part geometry. The tool nose radius (for the purpose of simplification only) in this example is programmed as if to be zero.

The following is an example using linear interpolation.

O3456

N001G50X10.Z5.S2500

N002T0200M41

N003G96S500M03

N004G00X1.35Z.2T0202M08

N005G01Z0F.01

N006X0

N007G00X1.11W.03

N008G01Z0

N009X1.25Z-.07

N010Z-1.5

N011X1.650

N012X1.75Z-1.55

N013X1.8

N014G00X10.Z5.T0200M05

N015M30

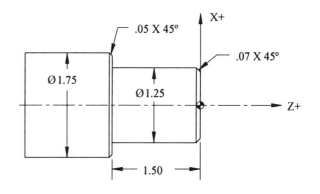

Figure 15

Circular Interpolation (G02 and G03)

Circular interpolation allows programmed tool movements along an arc. In order to define the circular interpolation function, the following conditions must be fulfilled.

1. Selection of the direction of interpolation:

G02 = Clockwise   G03 = Counterclockwise

Part 3 Preparatory Functions

2. Position coordinates of the ending point of the arc. The ending point coordinates may be omitted, if they correspond to the coordinates of the starting point(half circle, circle).

3. The dimension corresponding to the distance between the center of the tool nose radius and the center of the arc, performed from the starting point of each axis must be given.

The distance for the X axis is defined by the value of letter address I, and the Z axis is defined by the value of the letter address K. Values for I and K may be omitted from the program, if they are equal to zero. The block format for circular interpolation is given as follows.

G02X(U) ... Z(W) ... I ... K ... F ...

G03X(U) ... Z(W) ... I ... K ... F ...

The following figure graphically identifies all of the components necessary for programming arcs and their descriptions.

A = Arc starting point

B = Arc ending point

I = Incremental distance to the arc center from along the X axis

K = Incremental distance to the arc center from along the Z axis

O = Arc center

R = Arc radius

U/2 = Incremental distance from the arc starting point to the ending point along the X axis

W = Incremental distance to the arc end point along the Z axis

X/2 = Absolute coordinate for the ending point of the arc along the X axis

Z = Absolute coordinate for the ending point of the arc along the Z axis

In order to establish signs for *I* and *K*, consider the following directions: Imagine a vector is drawn from the arc starting point to the arc center point with a direction vector toward the center of the arc. Next, project this vector onto the axes of the coordinate system. If the resulting projections of the vector are oriented in the same direction as the corresponding axis of the absolute coordinate system the sign plus (+) is applicable. Otherwise, the sign minus (-) is applicable. If no sign is given in the coordinate entry, the control assumes the sign is positive.

The following figure is a pictorial representation showing how to determine the sign for vectors of *I* and *K*, while taking into account the tool nose radius.

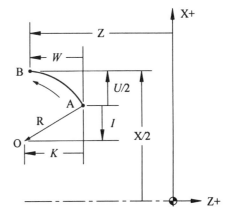

Figure 16

Part 3 Preparatory Functions

Figure 17

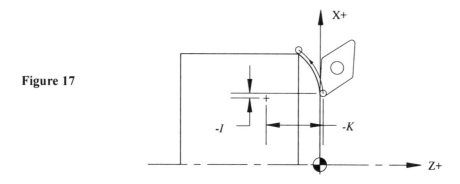

The following Figure 18 shows how to determine the direction of the circular interpolation and signs for vectors *I* and *K*. (The tool tip in the drawing is represented by a circle for easier pictorial presentation.)

Figure 18

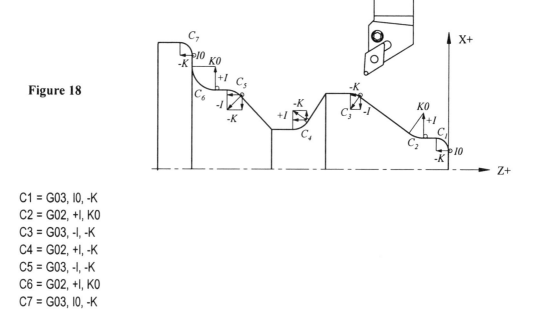

C1 = G03, I0, -K
C2 = G02, +I, K0
C3 = G03, -I, -K
C4 = G02, +I, -K
C5 = G03, -I, -K
C6 = G02, +I, K0
C7 = G03, I0, -K

Modern CNC controls include an additional capability to use *R* in place of *I*, *J*, and *K*. *R* is the distance from the center of the tool radius to the center of the following arc. If an arc is smaller than or equal to 180°, then *R* assumes a positive sign; if it is greater than 180°, then *R* assumes a negative sign. The block format for circular interpolation using R is given as follows.

   G02X(U)...Z(W)...R... F...

   G03X(U)...Z(W)...R... F...

Circular interpolation can be performed two different ways. The application of address R, however, is less difficult.

Part 3 Preparatory Functions

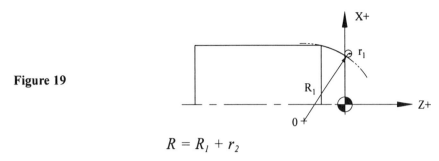

Figure 19

$$R = R_1 + r_2$$

If *I* and *K* are applied, then machine control is fed with precise information about the position of the center of the radius of the performed arc. In this case, the coordinates of the arc ending point must correspond to a position on the programmed circle. If, however, the given values of the coordinates are incorrect, the tool will not respond by following an arc. If the second method using R is employed, the tool will follow an arc, even if the values of the coordinates are incorrect. (This is true, of course, if the ending point of the arc falls within the area of the diameter of the circle.)

The following is an example of circular interpolation utilizing I and K. Once again, in this example, only a finish cutting pass is programmed along the part geometry. The tool nose radius (for the purpose of simplification only) in this example is programmed as if to be zero.

Example of Program 2

O0002

N1G50X15.Z5.S2000

N2T0100M41

N3G96S500M03

N4G00X1.6Z.2T0101M08

N5G01Z0F.02

N6X0

N7G00X1.3W.03

N8G01Z0

N9G03X1.5Z-.1K-.1I0

N10G01Z-0.94

N11G02U.12Z-1.0K0.I.06

N12G01X2.2

N13G03X2.5W-.15I0K-.15

N14G00X15.Z5.T0100M05

N15M30

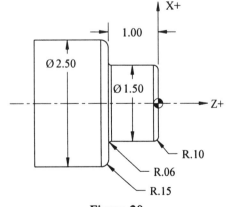

Figure 20

Part 3 Preparatory Functions

The following is the same example as above using R:

O0002

N1G50X15.Z5.S2000

N2T0100M41

N3G96S500M03

N4G00X1.6Z.2T0101M08

N5G01Z0.F.02

N6X0

N7G00X1.3W.03

N8G01Z0

N9G03X1.5Z-1.0R.1

N10G01Z-0.94;

N11G02U.12Z-1.0R.06

N12G01X2.2

N13G03X2.5W-.15R.15

N14G00X15.Z5.T0100M05

N15M30

Dwell (G04)

Dwell is initiated by use of function G04, and the length of time for the dwell is specified by P, X or U, (depending on the control type) as follows:

G04P . . . (in milliseconds)

G04X . . . (in milliseconds)

G04P . . . (in milliseconds)

Examples:

G04P2500

G04X2.5

G04U2.5

In the examples above, the dwell time values are the equivalent of 2 and 1/2 seconds. Also note that when using P to address the amount of time for dwell, a decimal point may not be used. The value of time is measured in milliseconds (ms), 1000 ms = 1 second. Function G04 is a "one shot" command and is active only in the block in which it is called. The dwell is activated at the end of the feed move and should be the only contents of the block. Dwell is sometimes indicated by the number of revolutions as opposed

Part 3 Preparatory Functions

to the amount of time (parameter setting). Study the manufacturer programming manual specific to the equipment to be sure of the exact method used. A common use for dwell is in the process of machining internal or external grooves, as shown in the following figure and described in the next paragraph.

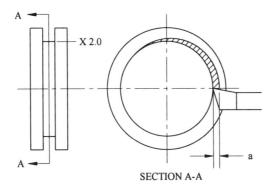

Figure 21

Examples:

G01X2.0F.008

G04U.25 (or G04X0.25 or G04P250)

G00X2.5

In the process of making the groove, you must remove a layer of material with a thickness corresponding to the depth of the cut and equivalent to the feed per revolution, as the tool tip reaches the diameter indicated in the program, in order to avoid an egg-shaped workpiece. If function G00 follows function G01, the resulting shape of the workpiece will be that of an egg because the tool is removed from the groove before all of the material can be cut. In the above example, the programmed dwell is 1/4 second and this allows a sufficient pause necessary to clean up the bottom of the groove before retracting from it.

Automatic Reference Point Return Check (G27)

Function G27 is an automatic return function. This command positions the tool, at rapid traverse, to a reference position indicated by the coordinates of X (U) and Z (W) in the block. It serves the purpose of confirmation of the return to the reference position. The reference position does not have to be a position of machine zero, but it commonly is, because of the safety it insures. The block format for automatic reference point return check (G27) is given as follows.

G27X(U)...Z(W)...T...00

Application of this function is shown in the following program example:

Part 3 Preparatory Functions

Example of Program 3

```
O0003
N1G50X15.Z5.0S800
N2T0100M41
N3G96S500M03
N4G00X1.0Z.1T0101M08
N5G01Z0F.01
N6X0
N7G00X.9Z.03
N8G01Z-1.0
N9G27X15.Z5.0T0100M05
N10M30
```

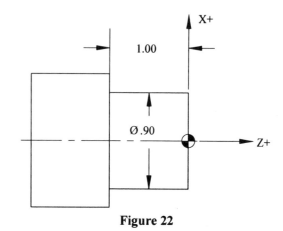

Figure 22

Notes: 1. The block that contains function G27 must also include cancellation of the wear offset.

2. Function G27 is only valid for function G50's axis input.

3. The time delay for the confirmation of the reference point return check by the control is approximately between 0 and 3 seconds.

4. A correct return to the reference position will be confirmed by a visual inspection of the control LED's being lit for the given axes. If the return position does not check accurately, an alarm will be displayed and the machine will stop until the error is corrected.

At first glance, it may seem senseless to use this function; however, a closer look provides sufficient reasons. One good example is for tool nose radius compensation. If the programmer inputs functions G41 and G42, but for some reason omits cancellation of function G40, then the tool will return to its reference position with a displacement equal to the offset. Without the use of G27, in this case, the next cycle may result in the destruction of the workpiece.

Automatic Reference Point Return (G28)

One of the ways to command a tool to return to the reference point after completion of an operation is through the application of function G28. The block format for Automatic Reference Point Return (G28) is given as follows.

    G28X(U)...Z(W)...

In the above block, the values in X(U) and Z(W) are the coordinates for an intermediate point through which the tool will pass, on its way to machine zero return. The tool

Part 3 Preparatory Functions

will position at a rapid traverse rate in nonlinear form. Therefore, it is recommended that an escape move be programmed, so that the tool is clear of the part before commanding G28. When using G28 tool offsets should be cancelled in a prior block.

T0100 = Cancellation of tool offset

Example of Program 4

O0004
N1G50X10.Z8.0S800
N2T0100M41
N3G96S200M03
N4G00X.5Z.1T0101
N5G01Z-.4F.008
N6X.8
N7T0100
N8G28X.9Z.1M05
N9M30

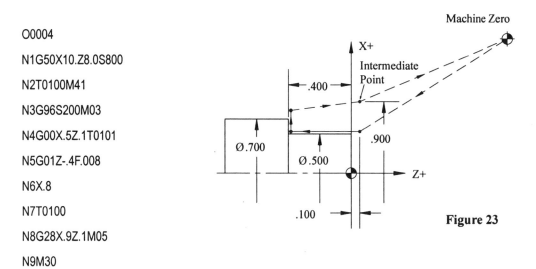

Figure 23

Block N8 is programmed in the absolute coordinate system. To change the command into the incremental system, follow this format:

N8G28U.1W.5M05

If no obstacles appear on the tool's path as the tool proceeds to machine zero return, then most often the intermediate point is merely a directional point.

N8G28X.8Z-.4M05 or

N8G28U0W0M05

Return From Reference Point (G29)

Function G29 generally follows G28 and must not be programmed without using G28 first. Using this command returns the tool, at rapid traverse, to a programmed point, by way of the same intermediate point as given in the G28 command. The block format used for G29 is as follows:

G29 X(U)... Z(W)...

N1G28U2.0W2.0

N2G29U-0.4W0.5

Part 3 Simple Cutting Cycles

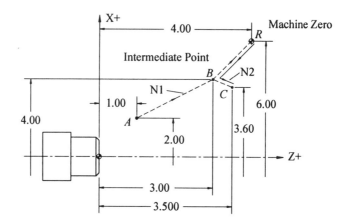

Figure 24

As shown in the drawing, the tool has moved from point A to point B (the intermediate point) and then on to point R (machine zero), all of which is determined in the first block with function G28. In block N2, the tool automatically returns from point R toward point B, and programming is limited to the calculation of the distance between points A and C. Point R is the machine zero position. Point C refers to the first position after tool change.

Thread Cutting (G32)

One of the oldest systems used (seldom) to program thread cutting is with the use of function G32. When it is used however, it offers absolute control of the threading if needed. The system uses a block-by-block programming technique to accomplish the desired thread. In other words, each pass of the threading tool must be programmed independently. The block format used for G32 is as follows:

1. Straight thread

   G32 Z(W)... F (or E ... for the Fanuc 6T control)

   where

   In the case of a straight thread, the diameter is determined by the previous block, which includes the X coordinate.

   Z(W) = Length of thread

   F, or E are equal to the thread pitch, 1/number of threads per inch.

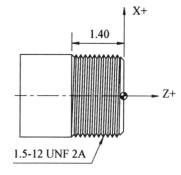

Figure 25

2. Taper thread

   G32 Z(W)... F (or E ... for the Fanuc 6T control)

   where

   X(U) refers to the dimension-determining diameter of the thread at the end points

Part 3 Simple Cutting Cycles

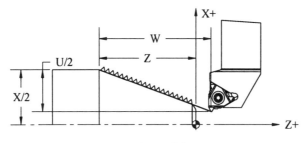

Figure 26

as in the case of a taper thread. In the case of a straight thread, X(U) may be omitted.

3. Face thread

    G32 Z(W)... F (or E ... for the Fanuc 6T control)

The feed rate for threading is programmed by using address F (or E and the value of feed are equivalent to the lead of the thread. Thread cutting using function G32 is rarely performed today because it is cumbersome to program each pass. Instead, function G92 is often used because it is easier to program, and function G76 is used most in these cases.

Threading Considerations

1. Changing the spindle speed and the feed rate override, while within the threading cycles, is not effective.
2. A "dry run" condition is applicable and effective.
3. The use of constant rotational speed programming, G97 S ..., is required!

Following is an example of straight thread cutting using function G32 for the part shown in Figure 25:

Example of Program 5

    O0005

    N1G50S1500

    N2T0100M41

    N3G97S600M03

    N4G00X1.58Z.5T0101M08

    N5G00X1.45

    N6G32Z-1.4F.0833

    N7G00X1.58

    N8Z.5

    N9X1.42

Part 3 Simple Cutting Cycles

N10G32Z-1.4

N11G00X1.58

N12Z.5

N9X1.4

N10G32Z-1.4

N11G00X1.58

N12Z.5

N13G28U0W0T0100M05

N14M30

Thread Cutting Cycle (G92)

By using G92 in one block of information, a cycle of four individual movements of the tool can be obtained. These movements are:

1. Rapid traverse to a given diameter.
2. Thread cutting with programmed feed rate.
3. Rapid withdrawal.
4. Rapid return traverse to the starting point.

Types of threads available are as follows.

1. Straight thread (cylindrical)

   G92X(U)...Z(W)...F or E...

2. Taper thread

   G92X(U)...Z(W)...I...F or E...

By activating SINGLE BLOCK, the introduced movements of a, b, c, and d can be executed by pressing CYCLE START.

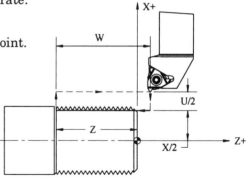

Figure 27

Note: By using function G92, it is possible to end the thread with a 45° chamfer or simply at a right angle.

As shown in Figure 29, a 90° thread end was obtained by using function M24, and a 45° chamfer is obtained by using function M23 as in Figure 30.

At the end of the threading routine, the G92 function must be cancelled by a G00 move. If it is not, the next programmed movement will continue as if it were still threading.

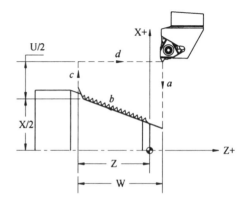

Figure 28

Part 3 Simple Cutting Cycles

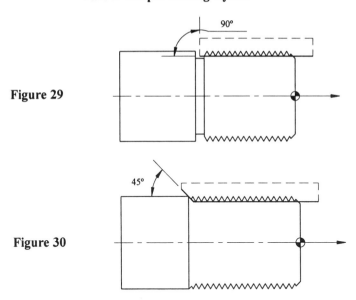

Figure 29

Figure 30

On some control models other than the Fanuc 16T, function G92 is a position register setting code. In these cases, when G92 is used for threading the G50 position register command must be used.

The previous notes related to threading using function G32 are also valid for G92.

The following is an example illustrating the application of function G92 for a straight thread:

Example of Program 6

O0006

N1G50S2000M41

N2T0100

N3G97S800M03

N4G00X1.1Z.5T0101M08

N5G92X.97Z-2.0F0.083333

N6X.95

N7X.92

N8X.90

N9G28U0W0T0100

N10M30

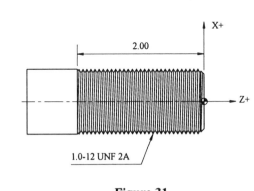

Figure 31

Note: Using this cycle with work requiring repetitive threading and cutting operations is very convenient, especially for large thread sizes. In such cases, after executing the block containing function G92, a certain number of consecutive blocks may be omitted and instead only the last few blocks may be executed.

Part 3 Simple Cutting Cycles

It is convenient to use function G92 in programs that result in much shorter length. For the above program, blocks N6, N7, and N8, deal with diameters for particular passes, while the preceding data that form block N5 are valid until block N9. The value of feed F (or E for some types of controls) may be expressed as shown in the preceding example with programmable accuracy reaching .000001 inch. In this example the value for Z in block N4 is .5. As far as the programmer is concerned, thread cutting may be initiated from a point positioned much closer to the material.

Threading could start as close as Z = 0. Practically speaking, however, it is not possible for the tool to begin the operation with the feed rate given in a program. Thus, part of the tool path is followed by the tool, with acceleration, until the tool reaches the value of the feed rate equivalent to thread lead. A similar situation occurs at the end of the threading process when the tool decelerates. A certain distance is traveled by the tool after some delay.

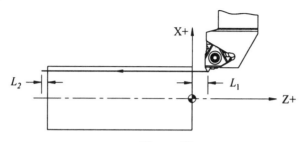

Figure 32

The lengths of L_1 and L_2, in the above figure, depend on the thread pitch and rotational speed of the spindle. Theoretical analysis of the mathematical formulas applied in calculating such lengths is rather complicated. For this reason, we will limit our explanation to the more simple calculations that are based on the following equations.

$$L_2 = \frac{P \times S}{1800} \text{ (in)}$$

where

P = lead (in)

S = revolutions per minute

$$L_1 = L_2 \times K \quad K = \ln\frac{1}{a} - 1 \quad a = \frac{\Delta P}{P}$$

where

ΔP = lead error

P = lead

The amount of lead error P depends on the design of the machine, as well as its servo motor systems. Values of the coefficient K, for a few specified error allowances, are contained in the table below:

Example:

Given and *P=0.0833, S=*1200, and a=0.01,

Part 3 Simple Cutting Cycles

$$L_2 = \frac{P \times S}{1800} = \frac{0.0833 \times 1200}{1800} = 0.0555$$

$$L_1 = L_2 \times K = 0.0555 \times 3.605 = 0.200$$

a	K
.02	2.91
.015	3.2
.01	3.605
.005	4.298

Chart 3

In most cases, these values are based on experience.

Tapered Thread Cutting using Cycle (G92)

When cutting a tapered thread using function G92 use the block format shown in item two above. The value of I is the radius or difference per side between the thread diameter at the end of the cut to the thread diameter at the start of the cut. Depending on the sign following the taper threading command of I, the cutting tool moves as follows:

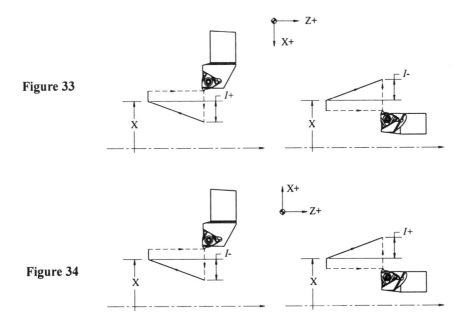

Figure 33

Figure 34

Fixed Cutting Cycle A (G90)

The fixed cutting cycle A is a cylindrical cutting function (to cut diameters). The block format used for G90 is as follows:

Part 3 Simple Cutting Cycles

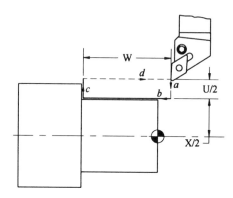

Figure 35

G90 X(U)...Z(W)...F...

Using function G90 in a program is a convenience. However, using function G90 will result in some loss of time because, after each pass, the tool returns by the motion d, as shown in the above figure.

The cycle execution is performed with four movements of the tool, as shown in the above figure, the same as it is with function G92.

Example of Program 7

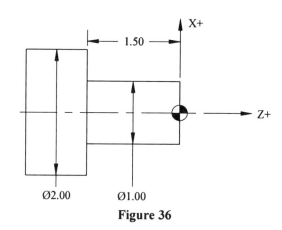

O0007

N1G50S2000M41

N2T0100M03

N3G96S500

N4G00X2.1Z.1T0101M08

N5G90X1.8Z-1.5F.015

N6X1.6

N7X1.4

N8X1.2

N9X1.0

N10G28U0.W0.T0100

N11M30

Figure 36

Note: Depending on the control type, tapered cuts may be programmed for function G90 by inclusion of letter address I or R. Where either are input, there is a radial value of the difference between the starting and ending diameters.

Fixed Cutting Cycle B (G94)

G94 X(U)...Z(W)...F...

Part 3 Simple Cutting Cycles

Cutting cycle B is a function similar in application to function G90. It is used for facing.

As with function G90, the tool always returns after each pass to the starting point by motion d, as shown in Fig. 37. This is one reason why both methods of programming are not widely used in practice today. Multiple repetitive cycles are a much better choice.

Example of Program 8

O0008
N1G50S2000M41
N2T0100M03
N3G96S500
N4G00X2.6Z.05T0101M08
N5G94X1.5Z-.1F.015
N6Z-.2
N7Z-.3
N8Z-.4
N9Z-.5
N10Z-.6
N11G28U0W0T0100
N12M30

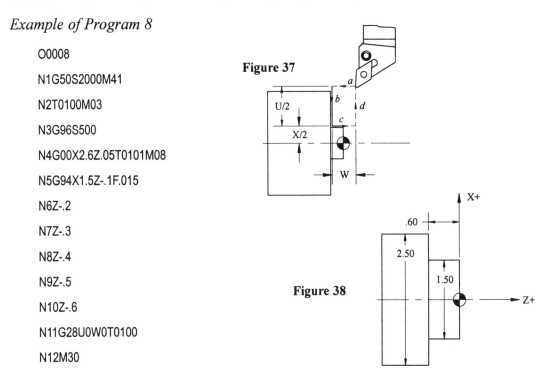

Figure 37

Figure 38

Note: Tapered cuts may be programmed for function G94 by inclusion of address K, where input is a radial value of the difference between the starting and ending diameters.

Multiple Repetitive Cycles

Rough Cutting Cycle (G71)

Function G71 is the rough cutting cycle that removes metal along the direction of the Z axis. In a case where there is a lot of material to be removed, this cycle provides an easy method for programming.

In the drawing Figure 39, the dotted lines refer to the initial shape of the workpiece, while the solid line refers to the final product. In the programming described so far, it has been necessary to employ many blocks of information to perform all the indi-

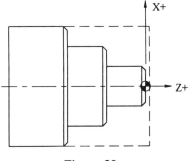

Figure 39

Part 3 Multiple Repetitive Cycles

vidual cuts for roughing. By using function G71, programming of the final shape of the workpiece is defined. Material is removed automatically, in each pass.

There are two types of program format for stock removal using function G71: single block and double block. The model of control used determines which type will be needed. Consult the manufacturer programming manual specific to the machine to determine the required method.

Function G71, Double Block

There are two program blocks required for function G71 when using this method. The finished profile of the part, as shown in Figure 40, is machined starting at point a, and proceeding to points b and c. The metal removal amount along the X axis is defined by the parameter U (depth of cut), in the first program block. A finish allowance for the X axis defined by the parameter U in the second block. Be careful not to get the two confused, as they do different things. The finish allowance for the Z axis is defined by W in the second block.

The following is a block diagram for function G71:

G71 U...R...

G71 P...Q...U...W...F...S...

where

in block one:

U = the depth of each roughing cut per side, to be used in consecutive passes (no sign)

R = the amount of retract, along the X axis, for each cut in block two

in block two:

P = the sequence number of the first block in the program, which defines the finish profile

Q = the sequence number of the last block of the program, which defines the finish profile

U = the stock allowance to be left for a finishing pass in the X axis direction (referred to as diameter; sign is + or -)

W = the stock allowance to be left for finishing in the Z axis direction (sign is + or -)

F = Cutting feed rate (in/rev or mm/rev) for blocks defined from P to Q

S = Spindle speed (ft/min or m/min) for blocks defined from P to Q

Figure 40

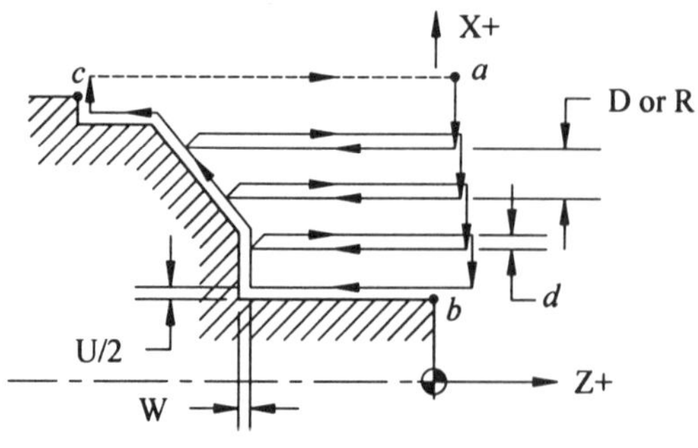

Part 3 Multiple Repetitive Cycles

The signs attached to symbols U and W may have negative or positive values, depending on the orientation of the coordinate system and the direction in which the allowance is assumed.

In the figure above, d represents the amount of X axis retract programmed for clearance by R in the first program block. This amount may also be set by parameter.

For the above figure:

a = the starting point of the given cycle

b = the sequence number of the first programmed point for the finish contour, which corresponds with P number of the second G71 block

c = the sequence number of the last programmed point for the finish contour

Notes for the double block command:

1. Changes in the feed between blocks P and Q will be ignored in G71. Only feed F, indicated by function G71, is valid.

2. The first tool path move of the programmed cycle from point a to point b cannot include any displacement in the direction of the Z axis.

3. The tool path between point b and c must be a steadily increasing, or decreasing pattern in both axes.

4. Both linear and circular interpolation is allowed.

5. The value for R must be noted as in the following examples:

R2000

R2500

R1500

6. For some controls, use of a decimal point may be allowed.

7. Figure 41 illustrates that if the allowance for finishing is located on the positive side (in the direction of the X and Z axes), with respect to the programmed contour, no sign is used; if on the negative side, the negative sign (-) is used.

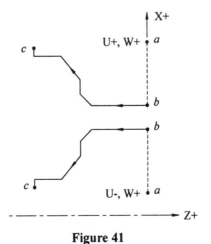

Figure 41

Part 3 Multiple Repetitive Cycles

Function G71, Single Block

There is only one program block required for function G71 when using this method. Much more freedom is allowed in regards to programmable shapes. In this case, it is not necessary to program a steadily increasing or decreasing pattern in both axes, it is only required along the Z axis, and up to ten concave figures are allowed.

Notable differences between double and single block function G71:

1. Block format for a single block is as follows:

   G71 P...Q...I...K...U...W...D...F...S...

where

All parameters are identical as stated above except:

I = Radial distance and direction of the rough cut along the X axis

K = Distance and direction of the rough cut along the Z axis

D = Depth of each roughing cut

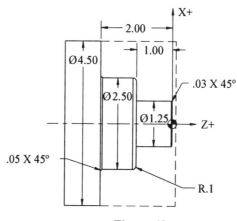

Figure 42

2. In the single block format, two axes may be programmed in the first block identified by the parameter P.

3. If single block format programming is used and the first block does not include any Z displacement, 0 must be input for parameter W in the G71 block.

Example of Program 9

The following is a program example for a double block call of G71:

O0009

N1G50S2000

N2T0100M41

N3G96S600M03

Part 3 Multiple Repetitive Cycles

N4G00X4.75Z.2T0101M08

N5G71U.12R.05

N6G71P7Q15U.03W.015F.019

N7G00X1.19

N8G01Z0

N9X1.25Z-.03

N10Z-1.0

N11X2.3

N12G02X2.5W-.1I0K-.1

N13G01Z-1.95

N14X2.6W-.05

N15X4.75

N16G28U0W0T0100M09

N17M05

N18M30

The following is a program example for a single block call of G71:

O0009

N1G50S2000

N2T0100M41

N3G96S600M03

N4G00X4.75Z.2T0101M08

N5G71P6Q15U.03W.003D1200F0.019

N6G00X1.19

N7G01Z0

N8X1.25Z-.03

N9Z-1.0

N10X2.3

N11G02X2.5W-.1I0K-.1

N12G01Z-1.95

N14X2.6W-.05

N15X4.75

N16G28U0W0T0100M09

Part 3 Multiple Repetitive Cycles

N017M05

N018M30

Comments: The block N6 first appears right after the block containing function G71. Despite the fact that the last programmed cycle block N15 refers to the location of tool tip Z-2.0, the tool will automatically return to the starting point indicated as Z.2 in the block N4.

Face Cutting Cycle (G72)

The properties for function G72 are similar to G71. The only difference, is in the change of the cutting direction to facing. The following is a function block diagram:

single block:

G72 P... Q ... I ... K ...U...W... D... F... S...

double block:

G72 W... R...

G72 P... Q... U... W... D... F... S...

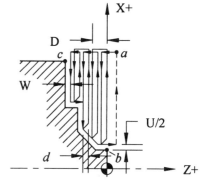

Figure 43

The parameters for this function have the same meaning as those described for function G71.

Notes: 1. The first block of the programmed cycle should not include any displacement in direction of the X axis. 2. The remaining notes for this function are identical to those for function G71. 3. The principle defining the choice between a positive or negative sign for U and W is identical to function G71.

Example of Program 10

O0010

N1G50S2000

N2T0100M41

N3G96S500M03

N4G00X4.4Z.2T0101M08

N5G72P6Q11U.03W.012D1000F.012

N6G00Z-1.5

N7G01X3.5

N8X3.1Z-1.3

N9X2.0

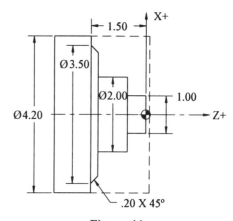

Figure 44

Part 3 Multiple Repetitive Cycles

N10Z-0.5

N11X1.0

N13G28U0W0T0100M09

N014M05

N015M30

Pattern Repeating (G73)

Function G73 permits the repeated cutting of a fixed pattern, with displacement of the axes position by an amount determined by the total material to be removed, divided by the number of passes desired. This cycle is well suited for previously formed castings, forgings or rough machined materials. This machining method assumes that an equal amount of material is to be removed from all surfaces. It can still be used if the amounts are not equal, but caution should be applied concerning excessive depths of cut and there may also be occasions of air cutting. Basically, the best scenario is when the finished contour closely matches the casting, forging, or rough material shape.

The following is a block diagram for function G73 using the double block format:

> G73 U...W...R...
>
> G73 P...Q...U...W...F...S...

where

In block one

U = total displacement in the direction of the X axis (sign + or -)

W = total displacement in the direction of the Z axis (sign + or -)

R = the number of rough cutting passes

In block two

P = number of the first block of the finished profile (given in the following figure as position "b")

Q = number of the last block of the finished profile (given in the following figure as position "c")

U = finish stock allowance in the direction of the X axis (sign + or -), referred to the diameter

W = finish stock allowance in the direction of the Z axis (sign + or -)

F = feed rate, effective for blocks P through Q

S = spindle speed, effective for blocks P through Q

Notes on function G73 using the double block format:

1. *Do not confuse the function of U and W with those in the single block format.*
2. *There is no use of address, D, in the double block format. Depth of cut is automatically calculated by the control based on U, and W, stock removal amount in*

Part 3 Multiple Repetitive Cycles

X and Z axes and R, the number of cutting passes.

The following is a block diagram for function G73 using a single block:

G73 P...Q...U...I...K...U...W...D...F...S...

where

P = number of the first block of the finished profile (given in the following figure as position "b")

Q = number of the last block of the finished profile (given in the following figure as position "c")

I = total displacement in the direction of the X axis (sign + or -)

K = total displacement in the direction of the Z axis (sign + or -)

U = finish stock allowance in the direction of the X axis (sign + or -), referred to stock left on the diameter

W = finish stock allowance in the direction of the Z axis (sign + or -)

D = the number of rough cutting passes

F = feed rate, effective for blocks P through Q

S = spindle speed, effective for blocks P through Q

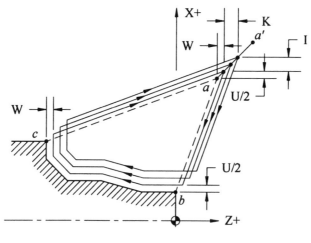

Figure 45

Point a on the drawing is the starting point. In executing this cycle, the tool travels from point a to point a', where displacements are defined by the values of I and K, as well as allowances for U and W, and, the cycle begins equivalent cutting passes with the number of passes, D. At the end of the cycle, the tool automatically returns to point a. Points b to c define the finished profile to be machined by function G70 as described in the next section.

Address I and K are measured on the machined workpiece. The principle defining choice of signs is similar to that for U and W.

Part 3 Multiple Repetitive Cycles

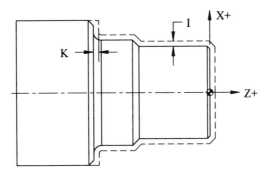

Figure 46

Example of Program 11

    O0011

    N1G50S2000

    N2T0100M41

    N3G96S500M03

    N4G00X2.2Z.3T0101M08

    N5G01Z.01F.03

    N6X0.F.012

    N7G00X3.0Z.2

    N8G73P9Q16I.168K.169U.04W.02D3F.012

    N9G00X1.59

    N10G01Z0

    N11X1.75Z-.08

    N12Z-1.375

    N13X2.0W-.125

    N14Z-2.1

    N15G03U.3Z-2.25I-.15K0.F.004

    N16G01X2.85

    N17G28U0W0T0100

    N18M30

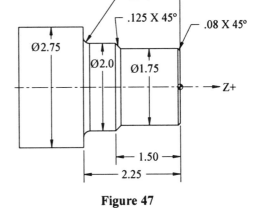

Figure 47

Finishing Cycle (G70)

Stock allowances for finishing (U, W) may be removed by the same tool used in rough cutting. However, it is a common practice to use a different tool for the finishing

Part 3 Multiple Repetitive Cycles

pass. Application of function G70 allows for removal of the remaining stock allowance (with the previously applied cycles G71, G72, and G73, without repetitive passes, along the contour).

The following is a block diagram for function G70:

$$G70\ P\ldots Q\ldots F\ldots S\ldots$$

where

P = number of the first block of the finished profile (given in the above figure 45 as position "b")

Q = number of the last block of the finished profile (given in the above figure 45 as position "c")

F = feed rate, effective for blocks P through Q

S = spindle speed, effective for blocks P through Q

Notice from the block diagram, that it is only necessary to enter position coordinates of the first (b) through last block (c) of the previous rough cycle, which define the finished profile. This will cause an automatic return to the earlier part of the program for the coordinates needed for the completion of the process removing allowances U and W).

```
N1G50S1000

     .........................................

     .........................................

N4G71P5Q16U.02W.01D1000F.014

     .........................................            F.04

     .........................................            F.06

     .........................................            F.05

N20M01

N21G50S1500

     .........................................

     .........................................

N25G70P5Q16

     .........................................

N28M30
```

Attention: In the above program values of the feed used in information blocks following function G71 are ignored for the roughing cycle, but they are valid for function G70.

Part 3 Multiple Repetitive Cycles

Example of Program 12

O0012

N1G50S2000

N2T0100M42

N3G96S600M03

N4G00X5.2Z.2T0101M08

N5G71P6Q12U.04W.005D1000F.012

N6G00X1.5

N7G01Z-.5F.010

N8X3.5F.008

N9Z-1.5F.010

N10X4.8F.009

N11X5.1Z-1.7F.004

N12X5.2F.010

N13G00X7.75Z8.75T0100

N14M01

N16T0200S2000M41

N17G96S700M03

N18G00X5.2Z.2T0202M09

N19G70P6Q12

N20G28U0W0T0200

N22M09

N23T0100M05

N24M30

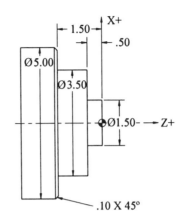

Figure 48

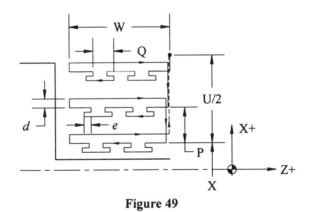

Figure 49

Peck Drilling Cycle (G74)

The most common use of this function is for drilling deep holes that require an interruption in the feed, in order to break long stringy chips. In spite of its name however, this function may be applied to cylindrical or face cutting of grooves that exhibit hard breaking chips as well. A block diagram of this function, as well as the movements of the tool, is illustrated abovein Figure 49.

Amount of the clearance (indicated by d in the above figure), is set by a system parameter. Amount of the return (indicated by e in the above figure), is also set by a parameter.

Part 3 Multiple Repetitive Cycles

The following is a block diagram for function G74 using a double block:

G74 R...

G74 X(U)...Z(W)...P...Q...R...F...S...

where

in the first block

R = retract amount of the tool after each cut

in the second block

X = diameter of the workpiece at the bottom of the groove

U = distance between the starting and end points (in an incremental system)

Z = final Z cut depth in the absolute system

W = Z distance from start to finish cut depth in the incremental system

P = depth of each cut for X axis (no sign)

Q = depth of each cut for Z axis (no sign)

R = retract amount of the tool at the bottom of the cutting

F = cutting feed rate

S = spindle speed

The following is a block diagram for function G74 using a single block:

G74 X(U)...Z(W)...I...K...D...F...S...

X = diameter of the workpiece at the bottom of the groove

U = distance between the starting and end points along the X axis (in an incremental system)

Z = final Z cut depth in the absolute system

W = Z distance from start to finish cut depth in the incremental system

I = depth of cut per side in X direction (no sign)

K = depth of cut per side in Z direction (no sign)

D = retract amount of the tool at the bottom of the cutting

F = cutting feed rate

S = spindle speed

In the following program example, this function is applied for drilling a deep hole using the single block format.

G74 Z(W)...K...F...

Example of Program 13

O0013

N1G50S1000

Part 3 Multiple Repetitive Cycles

```
N2T0100M41
N3G97S800M03
N4G00X0.Z.2T0101M08
N5G74Z-2.0K.550F.007
N6G28U0W0T0100M05
N7M30
```

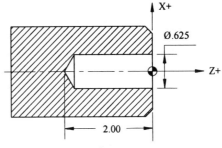

Figure 50

At the end of drilling (i.e., block N5), the tool automatically returns to the starting position of Z.2.

Advantages of the peck drilling cycle include:

1. chip breaking, and
2. cooling of the cutting drill tip.

Groove Cutting Cycle (G75)

Function G75, in its form, is very similar to function G74. Their difference lies in the direction of the tool movement (which is opposite that in function G74). Function G75 is used for cutting grooves that require an interrupted cut along the X axis.

The following is a block diagram for function G75 using a double block:

G75 R...

G75 X(U)... Z(W)... I... K... D... F... S...

The following is a block diagram for function G75 using a single block:

G75 X(U)... Z(W)... I... K... D... F... S...

All notations assumed for this function are defined exactly as in function G74.

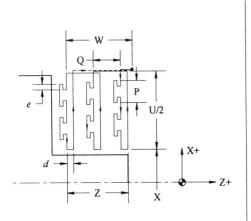

Figure 51

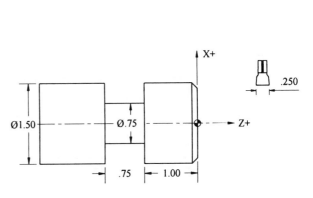

Figure 52

Part 3 Multiple Repetitive Cycles

Example of Program 14

O0014

(RIGHT SIDE OF INSERT CUTS)

N1G50S2000M41

N2G96T0100M03

N3G00X1.55Z-1.25S400M08

N4G75X.75Z-1.75T0101K.125D0F.005

N5G00X7.75Z8.5I.150M09

N6T0100M05

N8M30

Note: At the end of the cycle, the tool returns to the starting point in both functions G74 and G75.

Multiple Thread Cutting Cycle (G76)

The G76 multiple thread cutting cycle is used on most modern controls, in place of the outdated G32 and G92. All of the information needed to complete the desired thread is input in either one or two blocks, depending on the control, rather than multiple blocks in the former. By inputting the appropriate data for a particular type of thread in the program blocks, the number of cutting passes are automatically calculated by the control. Because of the limited number of blocks required, this method is very easy to edit.

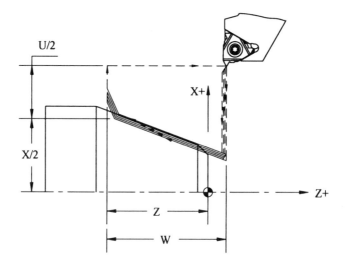

Figure 53

Part 3 Multiple Repetitive Cycles

The following is a block diagram for function G76 using a double block:

G76 P... Q... R...

G76 X... Z... R... P... Q... F...

where

in the first block

P = uses a six digit entry (P010000) of three pairs as follows:

the first two digits specify the number of finishing cuts (01 – 99)

three and four specifies the number of thread leads required for gradual pull-out (0.0 – 9.9 times the lead) without a decimal point entry (00 – 99)

five and six denote the angle of the thread (only 00, 29, 30, 55, 60 and 80 degrees are allowed)

Q = the minimum cutting depth

R = finishing allowance (allows a decimal point)

in the second block

X = Final diameter of the thread

Z = Final end position of the thread along the Z axis, (can be specified as an incremental distance using address W)

R = incremental distance from the thread starting to ending, as a radial value (used for tapered threads)

P = single thread height (always a positive radial value, without a decimal point)

Q = depth of cut for the first threading pass (always a positive radial value, without a decimal point)

F = feed rate, lead of thread

The following is a block diagram for function G76 using a single block:

G76X(U)... Z(W)... I... K... D... F... A... P...

where

X(U) = diameter of thread core (last diameter cut)

Z(W) = full length of the cut thread (end of thread position)

I = difference of thread radius (+ or -) from start to finish (for tapered threading)

K = single thread height (always positive)

D = depth of first threading cut (always positive)

A = thread angle (matches the included angle of the threading insert and is always positive)

F = feed rate, lead of thread

Notes:

1. All notes from functions G32 and G92 are applicable to function G76.

2. The depth of a first cut D is approximately .003 to .018, depending on machining conditions.

Part 3 Tool Nose Radius

3. *With a small value of first cut, the number of passes increases and, inversely, with a greater value of the first cut, the number of passes decreases.*

4. *The selection of factor D depends on the type of thread, the material, and the tool tip.*

Example of Program 14

O0014

N1G50S2000

N2T0100M41

N3G97S800M03

N4G00X1.35Z.5T0101M08

N5G76X1.2Z-1.5I0K.055D.012A60E083333

N6G28U0W0T0100M09

N7M30

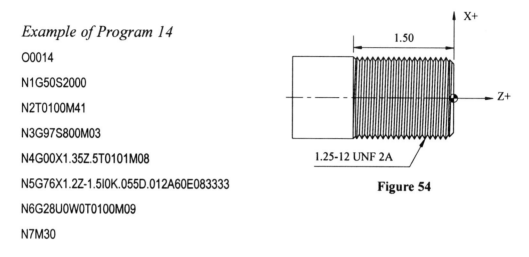

Figure 54

Note: *Depending on the value of the first cut specified by D, a certain number of passes will be obtained, as defined by the parameters that are set for the machine.*

Progamming for the Tool Nose Radius

In all of the programming examples examined thus far, the tool nose radius was considered to be zero. In reality, there is no such tool and the contour of the tool nose corresponds to the radius of its circle.

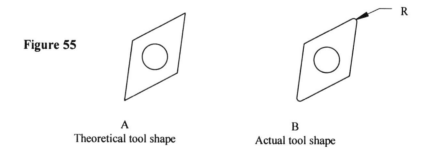

Figure 55

A
Theoretical tool shape

B
Actual tool shape

In the turning process on conventional lathes (in many cases), tool sharpening is performed on High Speed Steel (HSS) tool bits. On CNC lathes, HSS tools are rarely used or sharpened. Instead, indexable inserted tools tips are widely used. These (exchangeable) inserts are made from sintered carbides, and the chemical constitution of each is defined for a particular material in the form of cutting grades. The tool nose

Part 3 Tool Nose Radius

radii for inserts are standardized and can be ordered to meet the needs for most applications. The following are some of the most common radius sizes available.

Examples:

1/64 or .0156, 1/32 or .0312, 3/64 or .0469, 1/16 or .0625

When programming with a tool radius that is assumed to be zero, the imaginary tool point is programmed within a given coordinate system of X and Z. However, as mentioned previously, inserts are used whose cutting tip is assigned a certain radius. Considering this fact, more complicated calculations must now be employed, in order to position the tool to compensate for the radius. The following figures demonstrate the two conditions for programming the use of tool nose radius.

The Center of the Tool Nose Radius

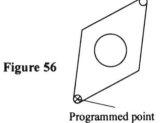

Figure 56

Programmed point

The Two Initial Points

Figure 57

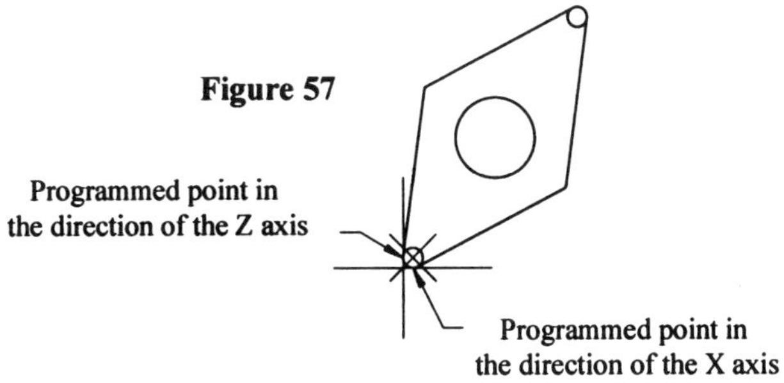

Programmed point in the direction of the Z axis

Programmed point in the direction of the X axis

In order to gain a better understanding of these conditions, examine the following example in which the tool nose radius, r = 0; then the above-mentioned cases can both be checked.

$$\tan 15° = \frac{.04}{a} \quad a = \frac{.04}{\tan 15°} = .1493$$

$$b = \frac{2.5 - (2.1 + 2 \times .1493)}{2} = .0507$$

$$\tan \beta = \frac{b}{.05} = \frac{.0507}{.05} = 1.0141 \quad \beta = 45.4°$$

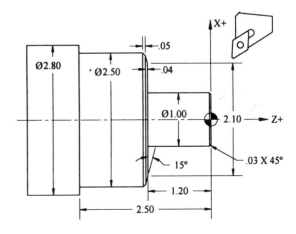

Figure 58

135

Part 3 Tool Nose Radius

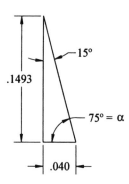 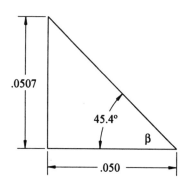

Figure 59

$\alpha = 75°$ $\beta = 45.4°$

Example of Program 15

O0015

N1G50S2000

N2T0100M41

N3G96S500M03

N4G00X1.1Z.2T0101M08

N5G01Z0.F.02

N6X0.F.01

N7G00X.940Z.03

N8G01Z0

N9X1.0Z-.03F.005

N10Z-1.2F.008

N11X2.1

N12X2.3986Z-1.24F.005

N13X2.5Z-1.29

N14Z-2.5F.008

N15X2.850

N16G00Z.5M09

N17G28U0W0T0100M05

N18M30

Part 3 Tool Nose Radius

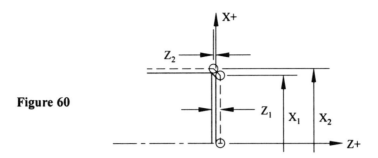

Figure 60

Programming the Center of the Tool Nose Radius

Before proceeding with programming of the center of the circle that describes the tool nose, examine a few cases where calculations are necessary. Please note that in the drawings for the remainder of this section, the tool nose representation has been enlarged in order to make dimensional data easier to see. The tool nose radius in these cases is .0312 inch and, is therefore difficult to see otherwise.

To program the .03 X 45° chamfer as the drawing illustrates, determination of the Z1 and X2 coordinates are relatively simple, while X1 and Z2 require additional calculations.

Case No. 1

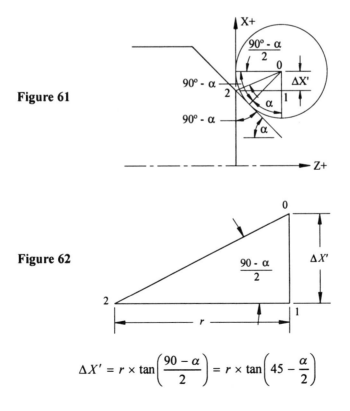

Figure 61

Figure 62

$$\Delta X' = r \times \tan\left(\frac{90 - \alpha}{2}\right) = r \times \tan\left(45 - \frac{\alpha}{2}\right)$$

Part 3 Tool Nose Radius

Case No. 2

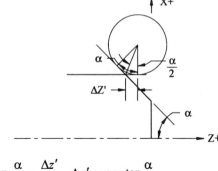

Figure 63

$$\tan\frac{\alpha}{2} = \frac{\Delta z'}{r} \quad \Delta z' = r \times \tan\frac{\alpha}{2}$$

Case No. 3

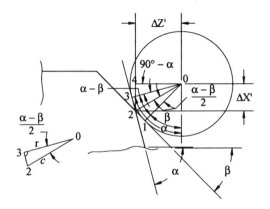

$$\cos\frac{\alpha - \beta}{2} = \frac{r}{c} \quad c = \frac{r}{\cos[(\alpha - \beta)/2]}$$

$$\gamma = 90 - (90 - \alpha) - \frac{\alpha - \beta}{2} = \alpha - \frac{\alpha}{2} + \frac{\beta}{2}$$

$$\gamma = \frac{\alpha}{2} + \frac{\beta}{2} = \frac{\alpha + \beta}{2}$$

Figure 64

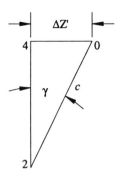

Figure 65

$$\sin\gamma = \frac{\Delta z'}{c} \quad z' = c \times \sin\gamma$$

$$\Delta z' = \frac{r}{\cos[(\alpha - \beta)/2]} \sin\frac{\alpha + \beta}{2}$$

$$\Delta z' = r\frac{\sin[(\alpha + \beta)/2]}{\cos[(\alpha - \beta)/2]}$$

Part 3 Tool Nose Radius

In the same manner, a formula for $\Delta X'$ can be derived

$$\Delta X' = r \frac{\cos[(\alpha + \beta)/2]}{\cos[(\alpha - \beta)/2]}$$

O0015

N1G50S2000

N2T0100M41

N3G96S500M03

N4G00X1.1Z.2T0101M08

N5G01Z.0312F.02

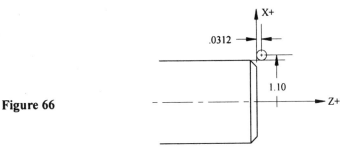

Figure 66

N6X0.F.01

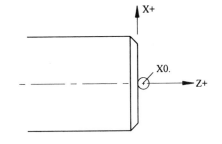

Figure 67

N7G00X.9658W.05

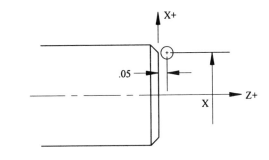

Figure 68

$$X = 1.0 - 2 \times .03 + 2 \times \Delta X' = 1.0 - .06 + 2\left[r \times \tan\left(45 - \frac{\alpha}{2}\right)\right] = .9658$$

Part 3 Tool Nose Radius

N8G01Z.0312

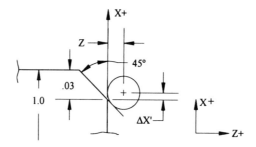

Figure 69

N9X1.0624Z-.0171F.005

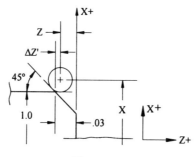

Figure 70

$$X = 1.00 + 2 \times r = 1.00 + 2 \times .0312 = 1.0624$$
$$Z = 0.03 - \Delta z' = .03 - r \times \tan\frac{\alpha}{2} = .03 - .0312 \times \tan\frac{45}{2} = .0171$$

N10Z-1.1688

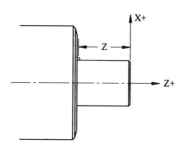

Figure 71

$$Z = 1.2 - r = 1.2 - .0312 = 1.1688$$

Part 3 Tool Nose Radius

N11X2.1082

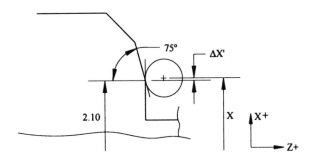

Figure 72

$$X = 2.1 + 2 \times \Delta x' = 2.1 + 2 \times \left[r \times \tan\left(45 - \frac{\alpha}{2}\right) \right]$$
$$= 2.1 \times \left[.0312 \times \tan\left(45 - \frac{75}{2}\right) \right] = 2.1082$$

N12X2.4310Z-1.212F.005

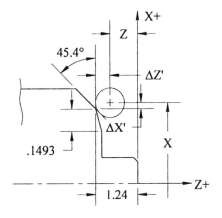

Figure 73

$$X = 2.1 + 2 \times .1493 + 2 \times \Delta X' = 2.1 + .2986 + 2 \times r \frac{\cos\left[(\alpha + \beta)/2\right]}{\cos\left[(\alpha - \beta)/2\right]}$$
$$= 2.3986 + 2 \times .0312 \frac{\cos\left[(75 + 45.4)/2\right]}{\cos\left[(75 - 45.4)/2\right]} = 2.4310$$
$$Z = 1.24 - \Delta Z' = 1.24 - r \times \frac{\sin\left[(\alpha + \beta)/2\right]}{\sin\left[(\alpha - \beta)/2\right]}$$
$$= 1.24 - .0312 \frac{\sin\left[(75 + 45.4)/2\right]}{\sin\left[(75 - 45.4)/2\right]} = 1.212$$

Part 3 Tool Nose Radius

N13X2.5624Z-1.2769

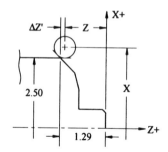

Figure 74

$X = 2.5 + 2 \times r = 2.5 + 2 \times .0312 = 2.5624$
$Z = 1.290 - \Delta z' = 1.290 - r \times \tan \alpha$
$Z = 1.290 - .0312 \times \tan 45.4° = 1.2769$

N14Z-2.4688

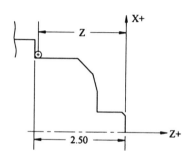

Figure 75

$Z = 2.5 - r = 2.5 - .0312 = 2.4688$

N15X2.9124

N16G00Z.5M09

N17G28U0W0T0100M05

N18M30

The last example leads to the following conclusions:

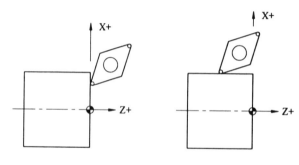

Figure 76

1. The operator will have difficulty in reading the program, because all the dimensions in the direction of the Z axis differ by the value of the tool radius, whereas, the diameters differ by double the value of the radius making changes virtually impossible.

2. In locating absolute zero (values of X and Z assigned to function G50), facing is performed with point No. 1 on the tool, whereas, the diameter is turned with point No. 2.

Part 3 Tool Nose Radius

It is clear, therefore, that to the resulting value of Z, the value of the tool radius must be added, whereas to the resulting value of X, double the value of the tool radius must be added.

Programming Using the Two Initial Points

The second method of programming refers to the description of the movements of a tool in two directions of motion: along the X and Z axes (assuming two different reference points).

Point No. 1 on the tool is a reference point in the direction of the Z axis, and point No. 2 on the tool is a reference point in the direction of the X axis. In order to make these changes, the formulas for X and Z must be rewritten as follows:

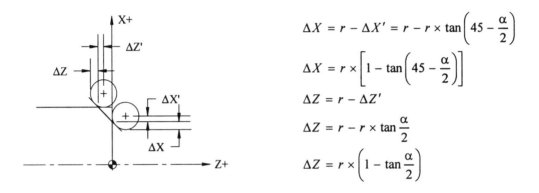

$$\Delta X = r - \Delta X' = r - r \times \tan\left(45 - \frac{\alpha}{2}\right)$$

$$\Delta X = r \times \left[1 - \tan\left(45 - \frac{\alpha}{2}\right)\right]$$

$$\Delta Z = r - \Delta Z'$$

$$\Delta Z = r - r \times \tan\frac{\alpha}{2}$$

$$\Delta Z = r \times \left(1 - \tan\frac{\alpha}{2}\right)$$

Figure 77

Rewrite the formulas for the double chamfer as follows:

$$\Delta X = r - \Delta X' = r - r\frac{\cos[(\alpha + \beta)/2]}{\cos[(\alpha - \beta)/2]} = r\left(1 - \frac{\cos[(\alpha + \beta)/2]}{\cos[(\alpha - \beta)/2]}\right)$$

$$\Delta Z = r - \Delta Z' = r - r\frac{\sin[(\alpha + \beta)/2]}{\cos[(\alpha - \beta)/2]} = r\left(1 - \frac{\sin[(\alpha + \beta)/2]}{\cos[(\alpha - \beta)/2]}\right)$$

O0015

N1G50S2000

N2T0100M41

N3G96S500M03

N4G00X1.1Z.2T0101M08

N5G01Z0F.02

Part 3 Tool Nose Radius

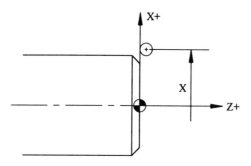

Figure 78

N6X-.0624F.01

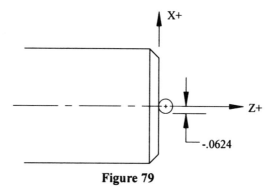

Figure 79

N7G00X.9036Z.03

N8Z0

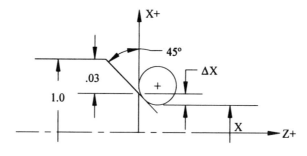

Figure 80

$$X = 1.0 - 2 \times .03 - 2 \times \Delta X = 1.0 - .06 - 2 \times r\left[1 - \tan\left(45 - \frac{\alpha}{2}\right)\right]$$

$$X = .94 - 2 \times .0312\left[1 - \tan\left(45 - \frac{45}{2}\right)\right] = .9036$$

Part 3 Tool Nose Radius

N9X1.0Z-.0483F.005

Figure 81

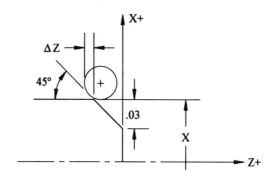

$X = 1.0$

$Z = .03 + \Delta Z = .03 + \left[r \times \left(1 - \tan\frac{\alpha}{2}\right)\right]$

$Z = .03 + \left[.0312 \times \left(1 - \tan\frac{45}{2}\right)\right] = .0483$

N10Z-1.2

Figure 82

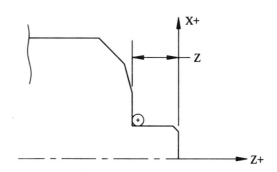

N11X2.0458

Figure 83

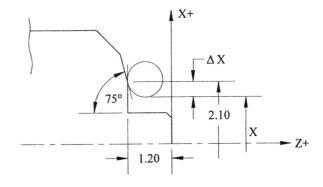

$X = 2.1 - 2 \times \Delta X = 2.1 - 2 \times \left[r \times \left(1 - \tan 45 - \frac{\alpha}{2}\right)\right]$

$X = 2.1 - 2 \times \left[1 - \tan\left(45 - \frac{75}{2}\right)\right] = 2.0458$

Part 3 Tool Nose Radius

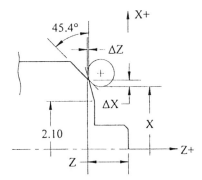

Figure 84

$$X = 2.1 + 2 \times .1493 - 2 \times \Delta X = 2.3986 - 2 \times r \left(1 - \frac{\cos\left[(\alpha + \beta)/2\right]}{\cos\left[(\alpha - \beta)/2\right]} \right)$$

$$X = 2.3986 - 2 \times .0312 \left(1 - \frac{\cos\left[(75 + 45.4)/2\right]}{\cos\left[(75 - 45.4)/2\right]} \right) = 2.3674$$

$$Z = 1.240 + \Delta Z = 1.240 + r \times \left(1 - \frac{\sin\left[(\alpha + \beta)/2\right]}{\cos\left[(\alpha - \beta)/2\right]} \right)$$

$$Z = 1.240 + .0312 \times \left(1 - \frac{\sin\left[(75 + 45.4)/2\right]}{\cos\left[(75 - 45.4)/2\right]} \right) = 1.2432$$

N12X2.3674Z-1.2432F.005

N13X2.5Z-1.3081

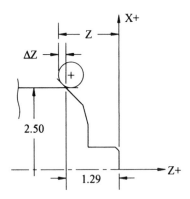

Figure 85

$$Z = 1.290 + \Delta z = 1.290 + r \times \left(1 - \tan\frac{\alpha}{2} \right)$$

$$Z = 1.290 + .0312 \times \left(1 - \tan\frac{45.4}{2} \right) = 1.3081$$

Part 3 Tool Nose Radius

N14Z-2.5F.008

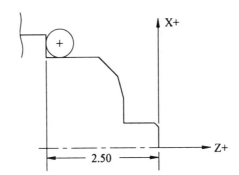

Figure 86

N15X2.85

N16G00Z.5M09

N17G28U0W0T0100M05

N18M30

Application of Tool Nose Radius Compensation (TNRC) G41, G42 and G40

Tool nose radius compensation is commonly used today. It adds ability to control the dimensional quality of geometry features, when using indexable inserts with a nose radius. It also aids in programming, by making it only necessary to program the workpiece profile without shifting the tool path to compensate for the tool nose radius. When a program is properly written using functions G41 or G42, setting the values correctly in the offset register of the control will produce a dimensionally accurate workpiece. The following explanations are given for the critical information needed for using functions G41 and G42.

Tool Nose Radius and Tip Orientation

In Figure 36 and 37 of Part II, Operation, the control display for tool geometry offset is shown. The last two columns in this register are used to input the values for the Tool Nose Radius (R) and the Tool Tip Orientation (T). These data are setup related, but have a very direct affect on the use of Tool Nose Radius Compensation (TNRC) in programming. If TNRC (G41 or G42) is used in the program, these data must be set accurately, or the programmed tool path will not generate the expected geometry. Straight facing or turning (parallel to either the X or Z axes) cuts do not require the use of TNRC, but in the case of tapered or circular contouring cuts and radii, TNRC is essential.

The following two figures detail the necessary information needed to select the proper setting for the Tool Tip Orientation (T), in the offset register, on a rear turret lathe.

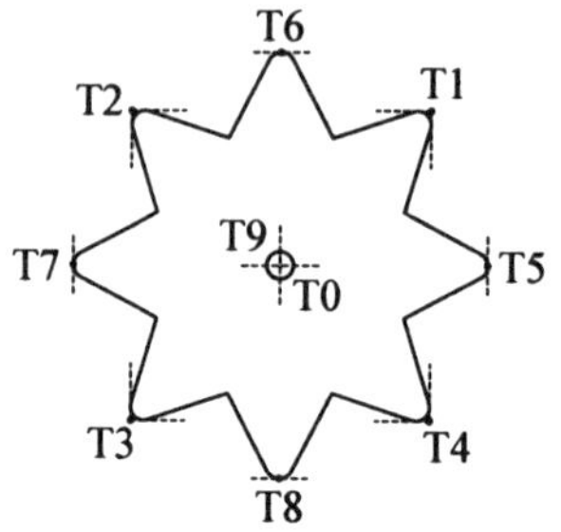

Part 3 Tool Nose Radius

Figure 87

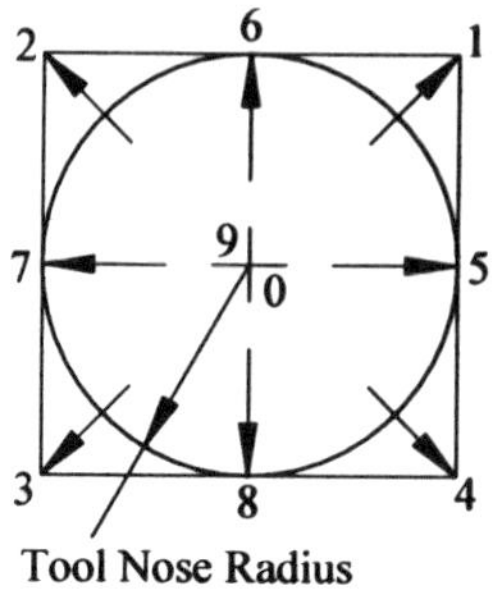

Figure 88

Tool Nose Radius

In the first figure, the tool tip is represented and the contact point is identified. The drawing should help with the selection of the T #, based on the direction the contact point is pointing. In the second figure, the same concept is superimposed onto the actual tool nose and the arrows indicate the contact point and direction.

Thus, the selection of the Tool Tip Orientation number is determined by the direction the imaginary tool nose is pointing. Use number 0 or 9 for programming the center of the tool nose.

Calling G41 or G42 in the Program

Selecting which code to use

Selection of either G41 or G42 is based on the side of the part profile that the tool needs to be on, to create the desired results. Think of the part profile as the centerline of a highway and, then, based on the direction of travel, decide which side of the road to drive on. For the left side, G41 is selected and, for the right side, G42 is used. This procedure may be applied whether the cut is internal, or external.

Initiating Cutter Compensation with G41 or G42

To initiate the use of tool nose radius compensation, the G41 or G42 should occur on a G00 rapid positioning move that is at least .100 of an inch away from the part profile. This move need only be in one axis direction, but it can include both. Please note how the G42 is initiated in the following example program for the part shown in Figure 58.

Ending use of G41 or G42 with function G40

In order to end the use of G41 or G42, TNRC may be cancelled by using function G40. When the machine is first started, the G40 command is active by default. To program the cancellation of tool nose compensation, the command is generally input on a move that is in departing vector from the machined profile. This move may be either G00 or G01 and cancellation may be initiated with the G28 command, where the compensation will be cancelled upon reaching the intermediate point. Please note how the G40 is used to cancel tool nose radius compensation in the following example program for the part shown in Figure 58.

Part 3 Tool Nose Radius

Example of Program 15 Using G41 or G42

```
O0015
N1G50S2000
N2T0100M41
N3G96S500M03
N4G00G41X1.1Z.2T0101M08
N5G01Z0.F.02
N6X-.04F.01
N7G00Z.03
N8G42X.940
N9G01Z0
N10X1.0Z-.03F.005
N11Z-1.2F.008
N12X2.1
N13X2.3984Z-1.24F.005
N14X2.5Z-1.29
N15Z-2.5F.008
N16X2.850
N17G00G40X4.Z.5M09
N18G28U0W0T0100M05
N19M30
```

Notes on using G41 and G42:

If the values for Tool Nose Radius (R) and the Tool Tip Orientation (T) are omitted in the offset register, the desired results will not be obtained (0 radius is assumed in this case). Note the change from G41 for facing, to G42 for the profile.

Programming Examples For Lathes

Description of Cutting Tools Used in Programming Examples

The following descriptions are for the tools used in the remaining examples:

Part 3 Programming Examples for Lathes

Tool number 01 = an outside diameter rough turning tool with an 80° diamond insert.

Figure 89

Tool number 02 = an outside diameter finish turning tool with a 55° diamond insert.

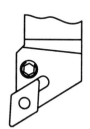

Figure 90

Tool number 03 = an inside diameter rough boring bar with an 80° diamond insert.

Figure 91

Tool number 04 = an inside diameter finish boring bar with a 55° diamond insert.

Figure 92

Tool number 05 = an outside diameter, U style, grooving tool.

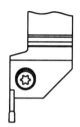

Figure 93

Tool number 06 = an inside diameter, U style, grooving tool.

Figure 94

Part 3 Programming Examples for Lathes

Tool number 07 = an outside diameter 60° threading tool.

Figure 95

Tool number 08 = an inside diameter 60° threading tool.

Figure 96

Tool number 09 = a #5 High Speed Steel center drill.

Figure 97

Tool number 10 = a standard drill.

Figure 98

Tool number 11 = a drill with triangular carbide inserts.

Figure 99

Tool number 12 = a reamer.

Figure 100

Tool number 13 = a tap.

Figure 101

Part 3 Programming Examples for Lathes

Application For Functions G00 And G01 In Both Absolute And Incremental Systems

Preliminary Considerations:

1. For the following example, the workpiece is rough turned with an allowance given for finishing; therefore, the use of only one finishing tool is programmed.
2. The part profile is programmed as if the tool nose radius is equal to zero.
3. The left side of the workpiece was faced in a previous operation.

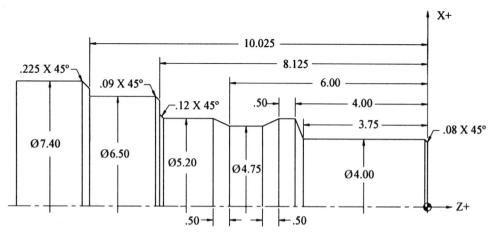

Figure 102

Example of Program 16 using an Absolute System

O0016

N1G50S1500

N2T0100M41

N3G96S600M03

N4G00X4.25Z.1T0101M08

N5G01Z0.F.03

N6X-.05F.01

N7G00Z.03

N8X3.84

N9G01Z0

N10X4.0Z-.08

N11Z-3.75

Part 3 Programming Examples for Lathes

N12X5.2Z-4.0

N13Z-4.5

N14X4.75Z-5.0

N15Z-6.0

N16X5.2Z-6.5

N17Z-8.005

N18X5.44Z-8.125

N19X6.32

N20X6.5Z-8.215

N21Z-10.025

N22X6.95

N23X7.5Z-10.35

N24G00Z.1M09

N25G28U0W0T0100M05

N26M30

Example of Program 16 using an Incremental System

O0016

N1G50S1500

N2T0100M41

N3G96S600M03

N4G00X4.25Z.1T0101M08

N5G01W-.1F.03

N6U-4.3F.01

N7G00W.03

N8U3.89

N9G01W-.03

N10U.16W-.08

N11W-3.67

N121U1.2W-.25

N13W-.5

N14U-.45W-.5

Part 3 Programming Examples for Lathes

N15W-1.0

N16U.45W-.5

N17W-1.505

N18U.24W-.12

N19U.88

N20U.18W-.09

N21W-1.81

N22U.45

N23U.55W-.325

N24G00U12.5W10.4M09

N25W14.9T0100M05

N26M30

Example of Program 16 using Absolute and Incremental Systems Combined

O0016

N1G50S1500

N2T0100M41

N3G96S600M03

N4G00X4.25Z.1T0101M08

N5G01Z0.F.03

N6X0.F.01

N7G00X3.84W.03

N8G01Z0

N9X4.0W-.08

N10Z-3.75

N11X5.2W-.25

N12Z-4.5

N13X4.75W-.5

N14Z-6.0

N15X5.2W-.500

N16Z-8.005

N17U.24W-.12

Part 3 Programming Examples for Lathes

N18X6.32

N19X6.5W-.09

N20Z-10.025

N21X6.95

N22X7.5W-.325

N23G00Z.1M09

N24X20.0Z15.T0100M05

N25M30

Eliminating Taper In Turning

Although CNC lathes are superior in accuracy to manually operated lathes, CNC lathes are far from perfect. Older machines are adversely affected by temperature changes, which in turn may influence the performance of electronic circuits. In addition, the temperature of the coolant may change the dimensions of turned parts. Even tool wear can cause similar variations. Such influential factors, during cylindrical turning, may cause differences in the values of the diameters for two points of the same part, as shown in the following figure. In the following example, tool # 1 (shown in Figure 89) is used and the insert has a .03125 nose radius.

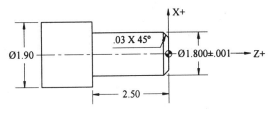

Figure 103

Example of Program 17

O0017

N1G50S1800

N2T0100M41

N3G96S600M03

N4G00G41X2.Z.1T0101M08

N5G01Z.0F.03

N6X-.065F.008

N7W.1

N8G00G42X1.7036W.03

Part 3 Programming Examples for Lathes

```
N9G01Z0
N10X1.8W-.0482
N11Z-2.5
N12X2.0
N13G00Z.2M09
N14G28U0W0T0100M05
N15M30
```

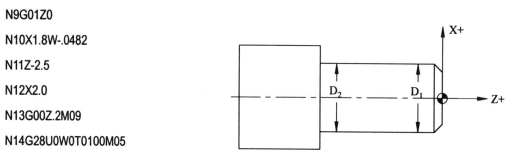

Figure 104

The measured diameters for points D_1 and D_2 are

$$D_1 = 1.8 \quad D_2 = 1.802$$

The correction offset can be assigned to block N11 as follows:

N11U-.002Z-2.5

For $D_1 = 1.8,$ $D_2 = 1.798$

N11U.002Z-2.5

For $D_1 = 1.802,$ $D_2 = 1.8$

N11U.002Z-2.5

Another common method of eliminating the taper that is less error prone (because the operator is adjusting the offset page rather than the program) is to use two offsets for the same tool. In this case the value is adjusted in the X axis wear offset register. The initial positioning move uses the original offset called in block N4 (T0101).

The correction wear offset can be assigned to block N11 as follows:

N11T0109Z-2.5

For $D_1 = 1.8,$ $D_2 = 1.798$

In this case the offset amount in the #9 X axis wear compensation register needs to be a negative (-.002) to correct the error.

For $D_1 = 1.802,$ $D_2 = 1.8$

In this case the offset amount in the #9 X axis wear compensation register needs to be positive (.002) to correct the error.

After eliminating the taper, the diameter will be equal for the entire turned length. To reduce the diameter, change the value in the program for U or the wear offset in a negative direction by a value of -.002 in.

Part 3 Programming Examples for Lathes

Eliminating Taper In Threading

Just as the statement implies, the situation of taper in threading is very similar to that of turning, but the actual method for elimination differs as described below.

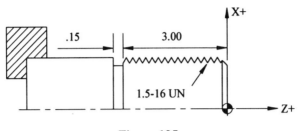

Figure 105

Example of Program 18

    O0018

    N1G50S1800M08

    N2T0700M41

    N3G97S764M03

    N4G00X1.6Z.5T0707M24

    N5G76X1.4234Z-3.075I0K.0383D01F0.0625

    N6G28U0W0T0700M23

    N7M09

    N8M30

As a reminder, look again at a part of the taper thread drawing, which first appeared as Figure 28 of this section.

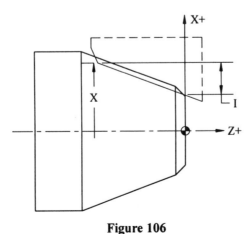

Figure 106

Part 3 Programming Examples for Lathes

Notice that X (the minor diameter) is the end point of threading, whereas magnitude I, appears at the beginning. By comparing this situation to straight turning, note another difference that results from the fact that I is calculated for one side only, unlike U.

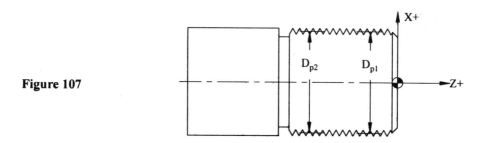

Figure 107

If the Dp_2 pitch diameter is correct, but Dp_1 is smaller by 0.003, then in block N5, assign .0015 for the value of I. If the Dp_2 pitch diameter is smaller than Dp_1 by 0.003, then I should read -.0015, and the wear offset value of .003 should be assigned in the positive direction for the X axis.

Subprogram Application

By using subprograms we can easily program parts that have identical repetitive features. An advantage of subprograms is that fewer lines of programming code are needed. The following is a simple example of how a subprogram might be used for cutting multiple grooves:

Example of Program 19

```
O0019
N1G501600
N2T0500M03
N3G96S400M41
N4G00X1.25Z-.375T0505M08
N5M98P0002L5
N6G28U0W0T0500M09
N7M30
```

Figure 108

Subprogram for Example of Program 19

```
O0002
N1W-.25
N2G01X1.0F.004
N3G00X1.25
N4M99
```

Part 3 Programming Examples for Lathes

In block N5 of the main program the subprogram O0002 is called and executed 5 times. When N4 of the subprogram is reached, the execution returns to the main program at block N6 and proceeds to its end.

Example Of Making A Taper Thread

For this example a 3/4 inch National Pipe Thread (NPT) will be programmed on a 4140 steel part. Before programming can begin, several calculations are necessary as shown below.

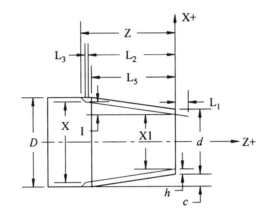

Figure 109

The thread dimensional data are drawn from Machinery's Handbook.

D = 1.05, diameter of pipe

L_2 = .5457, effective thread length

L_5 = .4029, length from the end of the pipe to the intersection of the taper and the outside diameter of the straight turned portion

h = .0571, height of the thread

Lead = 14

The angle of the thread is equivalent to 60°.

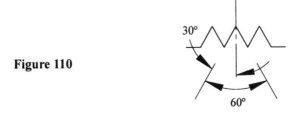

Figure 110

Taper of the thread on the diameter is 3/4 of an inch per foot.

Part 3 Programming Examples for Lathes

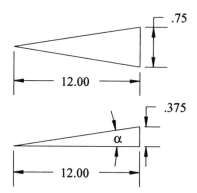

Figure 111

$$\tan \alpha = \frac{.375}{12.0} = .03125$$
$$\alpha = 1.7899° = 1°47'$$

Calculating diameter d, $d = D - 2c$

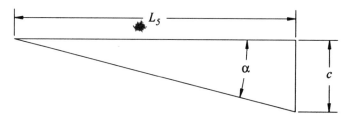

Figure 112

$$\tan \alpha = \frac{c}{L_5}$$
$$C = L_5 \times \tan \alpha = .4029 \times .03125 = .0126$$
$$d = 1.5 - 2 \times .0126 = 1.0248$$

The following is a calculation of the resulting length of "L", for the programmed section during threading.

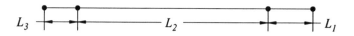

Figure 113

$$L = L_1 + L_2 + L_3$$
$$L_3 = \frac{S \times P}{1800} = \frac{727 \times .0714}{1800} = .028$$

160

Part 3 Programming Examples for Lathes

$a = 3.2$
$L_1 = L_3 \times a = .028 \times 3.2 = .0896 \approx .090$
$L = .09 + .5457 + .028 = .6637$

Calculating the value of I.

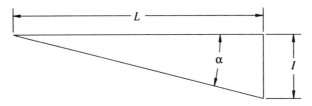

Figure 114

$$\tan \alpha = \frac{I}{L}$$
$I = L \times tg\,\alpha = .6637 \times .03125 = .0207$
$I = .0207$

Calculating the diameter of "X".

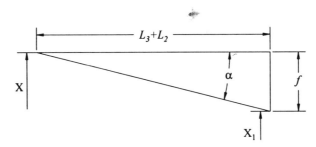

Figure 115

$X = X_1 + 2f$
$X_1 = d - 2h = 1.0248 = 2 \times .0571 = .9106$
$$\tan \alpha = \frac{f}{L_2 + L_3}$$
$f = (L_2 + L_3) \times \tan \alpha = (.5457 + .028) \times .03125 = .0179$
$X = X_1 + 2f = .9106 + 2 \times .0179 = .9464$
$X = .9464$

Example of Program 20

Note: In the following examples the setup sheet is used (as described in Part I of this text) because of the need for identifying multiple tools and other specific setup information.

Part 3 Programming Examples for Lathes

| Machine: Turning Center | Program Number: O0020 |

Workpiece Zero = X, <u>Centerline</u> Z, <u>Part Face</u>

Setup Description: Material = 4140 Alloy Steel AISI 1300

Tool # and Offset #	Tool Orientation #	Description	Insert Specification	Comments
T0101	3	O.D. Turning Tool	80° Diamond $\frac{1}{32}$ nose radius	400 SFM
T0707	8	O.D. Threading Tool	60° V Thread forming insert	300 SFM

Program

O0020

(OD TURNING TOOL)

N1G50S1500M08

N2T0101M42

N3G96S400M03

N4G00G41X1.2Z.2

N5G01Z0.F.03

N6X-.06F.008

N7G00W.03

N8G42X1.0266

N9G01Z0

N10X1.050Z-.4336F.007

N11G00U.06Z.2M09

N12G28G40U0W0T0100

N13M01

(OD THREADING TOOL)

N14G50S1000M08

N15T0707M42

N16G97S1000M03

N17G00G42X1.15Z.09

N18G76X.9464Z-.5737I-.0207K.0571A60D.012F.071428

N19G28G40U0W0T0700M09

N20M30

Notes: An offset number was introduced in block N2. In this case, the tool will advance rapidly to the position (X, Z), whose values are determined by the offset register. As described in previous comments, the resulting motion is defined as the movement to given coordinates with a displacement equivalent to the given values of the offsets.

Function G00 is used in block N4 and is valid until block N5 where it is replaced by a linear feed move.

Part 3 Programming Examples for Lathes

Example Of Turning With Bar Stock as the Material

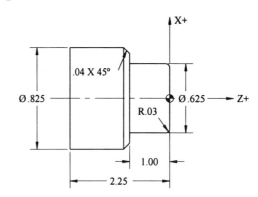

Figure 116

Example of Program 21

| Machine: Turning Center | Program Number: O0021 |

Workpiece Zero = X, <u>Centerline</u> Z, <u>Part Face</u>

Setup Description: Operation No. 1

Material: 1" diameter Carbon Steel 1100 bar stock

Tool # and Offset #	Tool Orientation #	Description	Insert Specification	Comments
T0101	3	O.D. Turning Tool	80° Diamond $\frac{1}{32}$ nose radius	600 SFM
T0505	3	O.D. Grooving Tool	.125 wide $\frac{1}{64}$ corner radius	300 SFM

Program

O0021

(OD TURNING TOOL)

N1G50S4000M08

N2T0100M42

N3G96S600M03

N4G00G41X1.1Z.2T0101

N5G01Z0F3.0

N6X-.06F7.0

N7G00W.030

N8G42X1.1Z.2

N9G71P10Q16U.03W.003D0800F0.016

N10X.565Z0

Part 3 Programming Examples for Lathes

N11G03X.625Z-.03I0K-.03

N12G01Z-1.0

N13X.745

N14X.825Z-1.04

N15Z-2.375

N16X1.1

N17G70P10Q16F.007

N18G28G40U0W0T0100

N19M01

(OD GROOVING TOOL)

N20G50S3000

N21T0500M42

N22G96S300M03

N23G00X1.1Z.2T0505M08

N24Z-2.3750

N25G01X.10F.012

N26X0

N27M05

N28G00X1.1

N29G28U0Z2.0T0500

N30M30

Example of Long Shaft Turning

The sequence of operations for turning the shaft in the Figure 117 are: (1) face and chamfer the shaft from the left side; (2) the program stops when using function M00; (3) turn the shaft around 180°; and (4) turn the remaining shaft end from the other side.

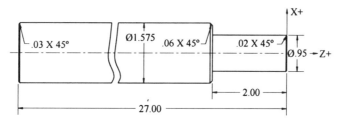

Figure 117

Part 3 Programming Examples for Lathes

Figure 118

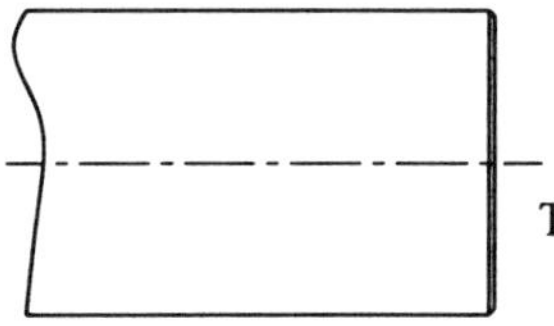

T01, T02
M00

Figure 119

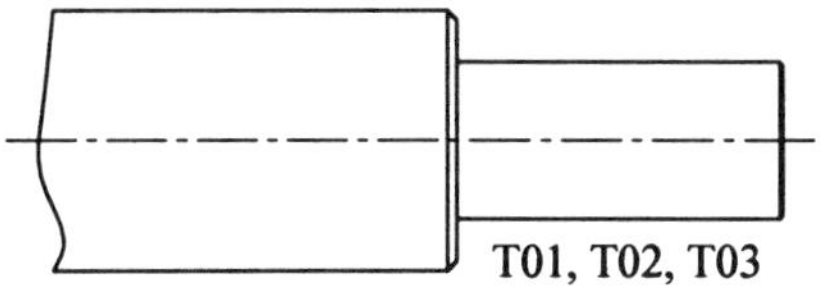

T01, T02, T03

Example of Program 22

Machine: Turning Center	Program Number: O0022

Workpiece Zero = X, <u>Centerline</u> Z, <u>Part Face</u>

Setup Description: Operation No. 1

Material: 1.625" diameter Tool Steel bar stock

Tool # and Offset #	Tool Orientation #	Description	Insert Specification	Comments
T0101 T0111	3	O.D. Turning Tool	80° Diamond $\frac{1}{32}$ nose radius	350 SFM
T0909		#5 HSS Center Drill		35 SFM
T0303	3	O.D. Turning Tool	55° Diamond $\frac{1}{32}$ nose radius	400 SFM

Program for the first operation

O0022

(ROUGH OD TURNING TOOL)

N1G50S1800

N2T0100M41

N3G96S350M03

N4G00G41X1.7Z.2T0101M08

N5G01Z0.F.02

N6X-.06F.01

N7G00W.03

N8G42X1.455

Part 3 Programming Examples for Lathes

N9G01X1.635W-.06F.004

N10G00Z.2M09

N11G28G40U0W0T0100

N12M01

(#5 CENTER DRILL)

N13G50S280M03

N14T0900M41

N15G00X0Z.1T0909M08

N17G01Z-.45F.006

N18G00Z.2M09

N19X25.0Z8.0T0900

N20M00

(ROUGH OD TURNING TOOL SECOND OFFSET)

N21G50S1800

At block No. N21, the value of Z is greater by .07 inch than the value of Z in block N1, because after the turning process of one end is finished, the shaft length was 27.07 inch; whereas, after the turning process of the other end is finished, the shaft length was 27.0 inches.

N22T0100M41

N23G96S350M03

N24G00G41X1.7Z.2T0111M08

N25G01Z.01F.02

N26X-.06F.014

N27G00W.03

N27G40X1.375

N28G01Z-1.99

N29G00U.06Z.1

N30X1.175

N31G01Z-1.99

N32G00U.06Z.1

N33X.975

N34G01Z-1.99

N35X1.7

Part 3 Programming Examples for Lathes

N36G00Z.2M09

N37G28U0W0T0100

N38M01

(FINISH OD TURNING TOOL)

In block N24, a new offset number (11) for tool number 1 is introduced, in order to maintain independent control of the length of both sides of the shaft.

N39G50S2000

N40T0300M41

N41G96S400M03

N42G00G41X1.1Z.1T0303M08

N43G01Z0.F.02

N44X-.06F.008

N45G00W.03

N46G42X.870

N47X.95Z-.02F.004

N48Z2.0F.008

N49X1.455

N50X1.695W-.120F.004

N51G00Z.2M09

N52G28G40U0W0T0300M05

N53T0100

N54M30

In block N53, tool No. 01 is called, in order to prepare this tool for the turning of the next operation.

The Second Operation

In the second operation, the cylindrical portion having a diameter equal to 1.575 inch is machined. The holding arrangement of the workpiece is shown below:

Figure 120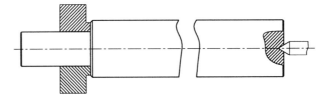

Part 3 Programming Examples for Lathes

Example of Program 23

Example of Program 23

Machine: Turning Center	Program Number: O0023

Workpiece Zero = X, <u>Centerline</u> Z, <u>Part Face</u>

Setup Description: Operation No. 2
Material: Tool Steel

Tool # and Offset #	Tool Orientation #	Description	Insert Specification	Comments
T0101	3	O.D. Turning Tool	80° Diamond $\frac{1}{32}$ nose radius	350 SFM
T0202	3	O.D. Turning Tool	55° Diamond $\frac{1}{32}$ nose radius	400 SFM

Program for the Second Operation

O0023

(ROUGH OD TURNING TOOL)

N1G50S1800

N2T0100M41

N3G96S350M03

N4G00X1.59Z.1T0101M08

N5G01Z-25.1F.012

N6G00U.06Z.1M09

N7G28U0W0T0100

N8M01

(FINISH OD TURNING TOOL)

N9G50S2000

N10T0200M41

N11G96S400M03

N12G00G42X1.6Z.1T0202M08

N13G01X1.575F.02

N14Z-25.1F.008

N15G00U.06Z.1M09

N16G28G40U0W0T0200M05

N17M30

Part 3 Programming Examples for Lathes

Programming Example For Making A Bushing

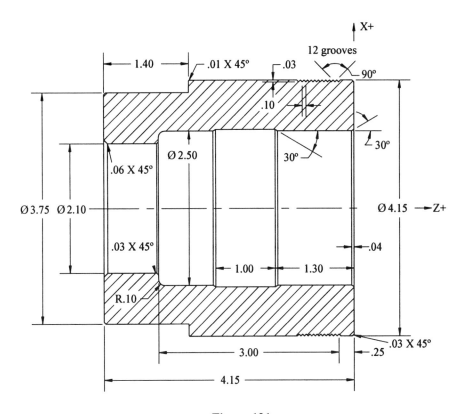

Figure 121

First operation

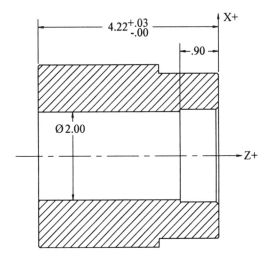

Figure 122

Part 3 Programming Examples for Lathes

In the first operation, the left side of the part including the hole with a 2.0 diameter and the length of 4.22 inch is machined. The length is finish machined in the next operation.

Example of Program 24

Machine: Turning Center	Program Number: O0024

Workpiece Zero = X, <u>Centerline</u> Z, <u>Part Face</u>

Setup Description: Operation No. 1

Material: Aluminum

Tool # and Offset #	Tool Orientation #	Description	Insert Specification	Comments
T0101	3	O.D. Turning Tool	80° Diamond $\frac{1}{32}$ nose radius	800 SFM
T0909		#5 HSS Center Drill		200 SFM
T1010		2.0 Diameter HSS Drill		200 SFM Minimum of 4.5 length
T0202	3	O.D. Turning Tool	55° Diamond $\frac{1}{32}$ nose radius	1200 SFM
T0303	8	I.D. BoringBar	80° Diamond $\frac{1}{32}$ nose radius	800 SFM

Program for the First Operation

O0024

(ROUGH OD TURNING TOOL)

N1G50S1000

N2T0100M41

N3G96S800M03

N4G00G41X4.4Z.2T0101M08

N5G01Z.02F.02

N6X-.06F.014

N7G00W.05

N8G00G40X4.W.05

N9G01Z-1.39

N10G00U.06Z.1

N11X3.78

N12Z-1.39

N13X4.35

Part 3 Programming Examples for Lathes

N14G28U0W0T0100M09

N15M01

(#5 CENTER DRILL)

N16T0900M42

N17G50S1830M03

N18G00X0.Z.2T0909M08

N19G01Z-.6F.006

N20G00Z1.M09

N21G28U0W0T0900

N22M01

(2-INCH DIAMETER DRILL)

N23T1000M41

N24G50S400M03

N25G00X0.Z.2T1010M08

N26G74Z-4.8K1.125F.008

N27G00Z1.M09

N28G28U0W0T1000

N29M01

(FINISH OD TURNING TOOL)

N30T0200M41

N31G96S1200M03

N32G00G41X3.85Z.1T0202M08

N33G01Z0.F.02

N34X1.8F.01

N35G00W.03

N36G42X3.63

N37X3.75W-.03F.004

N38Z-1.4F.001

N39X4.13

N40X4.15W-.01F.004

N41U-.03W-.03

Part 3 Programming Examples for Lathes

N42G28G40U0Z1.T0200M09

N43M01

(ROUGH ID BORING BAR)

N44T0300M41

N45G96S800M03

N46G00G42X-2.22Z.1T0303M08

N47X-2.1W-.06F.004

N48Z-1.

N49G00U.06Z1.M09

N50G28G40U0W0T0300M05

N51M30

Second Operation

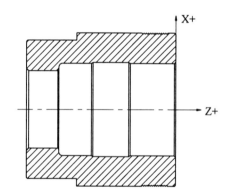

Figure 123

In the second operation, the remaining portion of the bushing will be machined from the opposite end.

Example of Program 25

Machine: Turning Center	Program Number: O0025

Workpiece Zero = X, Centerline Z, Part Face
Setup Description: Operation No. 2
Material: Aluminum

Tool # and Offset #	Tool Orientation #	Description	Insert Specification	Comments
T0101	3	O.D. Turning Tool	80° Diamond $\frac{1}{32}$ nose radius	800 SFM
T0303	8	I.D. Boring Bar	80° Diamond $\frac{1}{32}$ nose radius	800 SFM
T0202	3	O.D. Turning Tool	55° Diamond $\frac{1}{32}$ nose radius	1200 SFM
T0404	8	I.D. Boring Bar	55° Diamond $\frac{1}{32}$ nose radius	1200 SFM
T0707	8	Special 90° V-Form Tool		300 SFM

Part 3 Programming Examples for Lathes

Program for the Second Operation

O0025
(ROUGH OD TURNING TOOL)
N1T0100M41
N2G96S800M03
N3G00X4.4Z.2T0101M08
N4G01Z.02F.03
N5X1.8F.01
N6G00X4.16W.05
N7G01Z-2.85
N8G00U.06Z1.T0100M09
N9G28U0W0
N10M01
(ROUGH ID BORING BAR)
N11T0300M41
N12G96S800M03
N13G00X-2.2Z.2T0303M08
N14G01Z-3.24F.01
N15G00U.06Z.1
N16X-2.4
N17G01Z-3.24
N18G00U.06Z.1
N19X-2.47
N20G01Z-3.14
N21U.2Z-3.24
N22X-2.
N23G28U0Z.2T0300M09
N24M01
(FINISH OD TURNING TOOL)
N25T0200M42
N26G96S1200M03
N27G00G41X4.25Z.1T0202M08
N28G01Z0.F.03
N29X2.35F.008
N30G00W.03
N31G42G00X4.03
N32X4.15W-.03F.004
N33Z-2.85F.008
N34G00X4.5Z.1
N35G28G40U.0Z1.T0200M09
N36M01
(FINISH ID BORING BAR)
N37T0400M42
N38G96S1200M03
N39G00X-2.5723Z.1T0404M08
N40G01Z0F.012
N41X-2.5W-.0628F.004
N42Z-1.2963F.008
N43X-2.55W-.0433F.004
N44Z-2.3228F.008
N45X-2.5W-.0433F.004
N46Z-3.1812F.008
N47G03U.1376Z-3.25I.0688K0.F.003
N48G01X-2.1964F.008
N49X-2.05W-.0732F.004
N50G00Z1.M09
N51G28U0W0T0400
N52M01
(SPECIAL V-FORM TOOL)
N53T0700M41
N54G96S300M03

Part 3 Programming Examples for Lathes

N55G00X4.2Z-.25T0707M08

N56M98P5L12

N57G00Z1.M09

N58G28U0W0T0700M05

N59M30

Subprogram for Operation 2

O0005

N1G01X-4.09F.003

N2G00X-4.2

N3W-.1

N4M99

Notes for the second operation of the program

The following notes pertain to the calculations necessary for programming the external and internal chamfers and tapers, and indicate the specific block number where the resulting values are input. Once again, in the figures that follow, the tool nose radius representation is scaled up in order to aid in the visualization.

1. N39G00X-2.5723Z.1T0404M08

    X-2.5723 is the value that requires additional calculations.

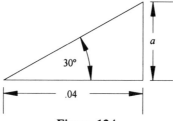

Figure 124

Based on the drawing, the given angle and the length of the chamfer enable us to calculate length a, of the triangle.

$$\tan 30° = \frac{a}{.04} \quad a = .04 \times \tan 30° = .0231$$

The expected value is given by a formula:

$$X = 2.5 + 2 \times a + 2 \times \Delta X$$

$$\Delta X = r \times \left[1 - \tan\left(45 - \frac{\alpha}{2}\right)\right]$$

where

Part 3 Programming Examples for Lathes

r = radius of the tool = $\frac{1}{32}$

a = acute angle of chamfer, with the symmetry axis at 30°

$$\Delta X = .0312 \times \left[1 - \tan\left(45 - \frac{30°}{2}\right)\right] = .0312 \times (1 - .5774)$$

$$= .0312 \times .4226 = .01318$$

$$X = 2.5 + 2 \times .02 + 2 \times .01318 = 2.5723$$

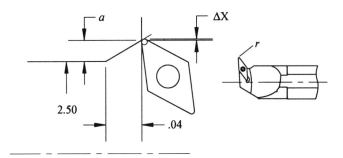

Figure 125

2. N41X- 2.5W-.0628F.004

Look at the displacement in the direction of the Z axis, represented by W-.0628.

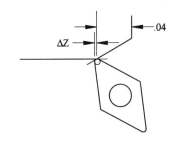

Figure 126

$$W = .04 + \Delta z$$

$$\Delta z = r \times \left(1 - \tan\frac{\alpha}{2}\right)$$

$$r = .0312 \quad \alpha = 30°$$

$$\Delta z = .0312 \times (1 - \tan 15°) = .02284$$

$$W = .04 + .0228 = .0628$$

3. N42Z-1.2963F.008

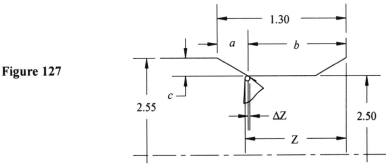

Figure 127

Part 3 Programming Examples for Lathes

Data, obtained from the drawing, refer to the dimensions 2.55, 2.5, and 1.3 and an angle of 30°.

$$z = b + 2 \times r - \Delta z$$
$$b = 1.3 - a$$

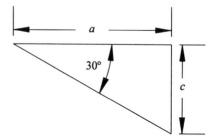

Figure 128

$$\tan 30° = \frac{c}{a} \quad c = \frac{2.55 - 2.5}{2} = .025$$

$$a = \frac{c}{\tan 30°} = \frac{.025}{\tan 30°} = .0433$$

$$b = 1.3 - a = 1.3 - .0433 = 1.2567$$

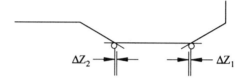

Figure 129

Since Z_1 is equal to Z_2, the same formula can be used:

$$\Delta z = r \times \left(1 - \tan \frac{\alpha}{2}\right)$$

$$\Delta z = .0312 \times \left(1 - \tan \frac{30}{2}\right) = .0228$$

$$z = b + 2 \times r - \Delta z = 1.2567 + .0624 - .0228$$
$$= 1.2963$$

4. N43X-2.55W-.0433F.004

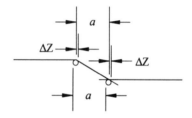

Figure 130

As the drawing indicates, the previously calculated chamfer length a is equal to the tool displacement in the direction of the Z axis in the process of making such a chamfer. A similar situation in block N48 occurs.

Part 3 Programming Examples for Lathes

5. N44Z-2.3228F.008

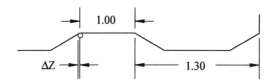

Figure 131

$$Z = 1.3 + 1.0 + \Delta z = 1.3 + 1.0 + .0228 = 2.3228$$

6. N46Z-3.1812F.008

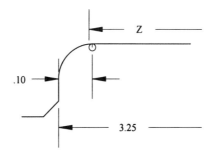

Figure 132

$$Z = 3.25 - .1 + .0312 = 3.1812$$

7. N47G03U.1376Z-3.25I.0688K0.F.003

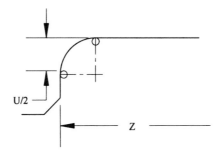

Figure 133

$$\frac{U}{2} = .1 - r \quad U = 2 \times (.1 - r) = .1376$$

Part 3 Programming Examples for Lathes

Example Illustrating The Application Of Functions G72 And G75

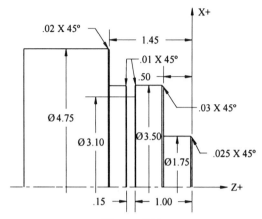

Figure 134

Example of Program 26

Machine: Turning Center	Program Number: O0026

Workpiece Zero = X, Centerline Z, Part Face
Setup Description: Operation No. 1
Material: 316 Stainless Steel

Tool # and Offset #	Tool Orientation #	Description	Insert Specification	Comments
T0101	3	O.D. Turning Tool	80° Diamond $\frac{1}{32}$ nose radius	250 SFM
T0202	3	O.D. Turning Tool	55° Diamond $\frac{1}{32}$ nose radius	400 SFM
T0505	8	O.D. Grooving Tool	.125 wide $\frac{1}{64}$ radius	250 SFM

Program

O0026

(ROUGH OD TURNING TOOL)

N1G50S1500

N2T0100M41

N3G96S250M03

N4G00G41X4.85Z.2T0101M08

N5G01Z.0F.04

N6X-.06F.012

N7G00Z.1

Part 3 Programming Examples for Lathes

N8G42X4.8
N9G72U.04W.005P9Q19D.12F.012
N10G00Z-1.47
N11G01X4.75
N12U-.02Z-1.45F.004
N13X3.5F.008
N14Z-.53
N15U-.03Z-.5F.004
N16X1.75F.008
N17G01Z-.025F.008
N18U-.025Z0F.004
N19Z.1
N20G28G40U0W0T0100M09
N21M01
(FINISH OD TURNING TOOL)
N22G50S2200
N23T0200M41
N24G96S400M03
N25G00G41X2.0Z.1T0202
N26G01Z0.F.04
N27X-.06F.008
N28G00Z.1
N29G42X4.8
N30G70P9Q19
N31G28G40U0W0T0200M09
N32M01
(OD GROOVING TOOL)
N33G50S2000
N34T0500M41
N35G96S250M03
N36G00X3.55Z.2T0505M08
N37Z-1.1375
N38G75X3.11I.05F.004
N39G00X3.55
N40Z-1.1591
N41G01X3.5F.008
N42U.0382W.0191F.003
N43X3.1F.005
N44W.005
N45G00X3.55
N46Z-1.1059T0313
N47G01X3.5F.008
N48U-.0382W-.0191F.003
N49X3.1F.005
N50W-.022
N51G00X3.75
N52Z1.M09
N53G28U0W0T0500
N54M30

In block N38, when the tool reaches diameter X3.11, it automatically returns to the starting point because of cycle G75. As a result of this, consecutive block N39 (with X3.55) is actually a safety clearance block since in some older types of controls, the automatic return may not appear. In cycle G75, some symbols such as K, D, and Z may be omitted, and, if any (K = 0, D = 0) are equal to zero, it may be omitted. The value of Z does not change and is omitted as well.

Part 3 Programming Examples for Lathes

Complex Program Example

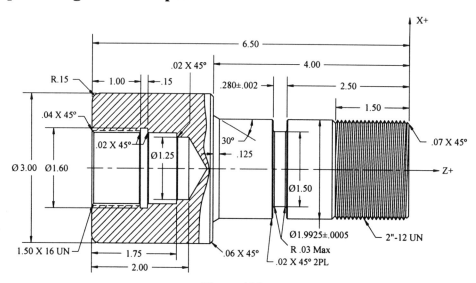

Figure 135

First Operation

1. The order calls for 1,025 pieces of material with a (3.125) diameter and a length of (6.625).

2. Complete all the dimensions on the left side of the part, along with a length of 6.60 inch, which leaves .100 inch for the finish pass on the second operation on the opposite end.

The following are the initial program data for the first operation:

1. The tool has a maximum diameter of .3 inch and the RPM for the center drill is calculated as follows:

Cutting speed for the steel is:
$$V = 80 \text{ ft/min}$$

RPM:
$$n = \frac{12 \times V}{\pi \times d} = \frac{12 \times 80}{3.14 \times .3} = 1019$$

The tool has a maximum diameter of 1.25 inch and the RPM for drill bit is calculated as follows:

$$V = 80 \text{ ft/min}$$

$$n = \frac{12 \times V}{\pi \times d} = 244$$

3. The cutting speeds for the boring bar, I.D groove, and the internal thread are V = 300 ft/min due to the unfavorable cutting conditions and the liklihood that vibrations will be present.

Part 3 Programming Examples for Lathes

4. The minor diameter for the 1.5-16 internal thread is obtained from the *Machinery's Handbook* and is from 1.432 minimum to 1.446 maximum.

Second Operation
Example of Program 27 Operation 1

Workpiece Zero = X, <u>Centerline</u> Z, <u>Part Face</u>
Setup Description: Operation No. 1
Material: Brass
Machine the left end including all internal work.

Tool # and Offset #	Tool Orientation #	Description	Insert Specification	Comments
T0101	3	O.D. Turning Tool	80° Diamond 1/32 nose radius	500SFM
T0909		#5 HSS Center drill		80SFM
T1010		1.25 diameter HSS drill		80SFM
T0303	2	I.D. Turning Tool	80° Diamond 1/32 nose radius	400SFM
T0202	3	O.D. Turning Tool	55° Diamond 1/32 nose radius	600SFM
T0404	2	I.D. Turning Tool	55° Diamond 1/32 nose radius	300SFM
T0606	2	I.D. Grooving Tool	.125 wide 1/64 radius	300SFM
T0808	2	I.D. Threading Tool	1/32 radius	300SFM

Program for the First Operation

O0027

(ROUGH OD TURNING TOOL)

N1G50S2000

N2T0100M41

N3G96S500M04

N4G00X3.2Z.2T0101M08

N5G01Z.02F.05

N6X-.0624F.014

N7G00X3.03W.03

N8G01Z-2.8

N9G00U.1Z1.0M09

N10G28U0W0T0100

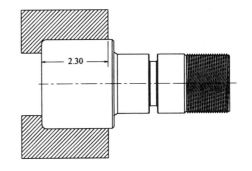

Figure 136

Part 3 Programming Examples for Lathes

N11M01

(#5 CENTER DRILL)

N12T0900M41

N13G97S1019M03

N14G00X0.Z.2T0909M08

N15G01Z-.3F.006

N16G00Z.5M09

N17G28U0W0Z10.0T0900

N18M01

(1.25 DIAMETER DRILL)

N19T1000M41

N20G97S244M03

N21G00X0.Z.1T1010M08

N22G74Z-2.360K.150F.008

N23G28U0W0T1000M09

N24M01

(ROUGH ID BORING BAR)

N25G50S2000

N26T0300M41

N27G96S400M04

N28G00X1.250Z.1T0303M08

N29G71U-.03W.005P30Q36D1000F.012

N30G00X1.5364

N31G01Z0

N32X1.420Z-.0582F.004

N33Z-1.75F.007

N34X1.3264

N35X1.25W-.0382F.004

N36G00Z.1M09

N37G28U0W0T0300

N38M01

(FINISH OD TURNING TOOL)

N39G50S2000

N40T0200M41

N41G96S600M04

N42G00X3.1Z.1T0202M08

N43G01Z0.F.05

N44X1.0F.008

N45G00X2.6376W.03

N46G01Z0

N47G02X3.0W-.1812F.005K-.1812

N48G01Z-2.8F.008

N49G00U.1Z1.0M09

N50G28U0W0T0200

N51M01

(FINISH ID BORING BAR)

N52G50S2000

N53T0400M41

N54G96S600M04

N55G00X1.250Z.1T0404M08

N56G70P30Q36

N57G28U0W0T0400M09

N58M01

(ID GROOVING TOOL)

N59G50S2000

N60T0600M41

N61G96S300M04

N62G00X1.4Z.2T0606M08

N63Z-1.1375

N64G01X1.590F.004

N65G00X1.4

N66Z-1.179

Part 3 Programming Examples for Lathes

N67G01X1.420F.008

N68U.058W.029F.004

N69X1.6F.005

N70W.005

N71G00X1.4

N72Z-1.096;

N73G01X1.420F.008

N74U.058W-.029F.004

N75G01X1.6F.005

N76W-.023F.006

N77G00X1.4

N78Z1.0M09

N87M30

N79G28U0W0T0600

N80M01

(ID THREADING TOOL)

N81T0800M41

N82G97S764M04

N83G00X1.340Z.5T0808M24

N84G76X1.5Z-1.075A60I0K.040D.0140F.0625

N85G28U0W0T0800M23

N86T0100M09

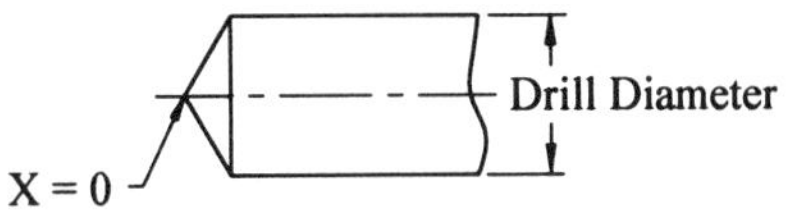

Figure 137

Notes: During the process of drilling, notice that function G97 is applied, which corresponds to a constant spindle speed. It is used in blocks N13 and N20, because when a centerline tool such as a drill is used, the assigned coordinate for its position is X0. In block N3 function G96, constant cutting speed is instated. If G96 were to remain active, the spindle speed for the X position of zero would calculate to infinity and the machine would be limited to its maximum programmable RPM. Obviously, this is not practical. In this program, the value of the spindle speed is limited, with the value of S assigned to function G50.

The drill bit, however, has a certain diameter. Therefore, the suitable RPM is calculated based on the diameter and entered in the program along with function G97.

When using function G96, there must be a programmed point located on the circumference in the direction of the X axis. This value, with respect to the X axis, must also be taken into consideration, if function G50 is assigned.

Part 3 Programming Examples for Lathes

Example of Program 27
Operation 2

| Machine: Turning Center | Program Number: O0027 |

Workpiece Zero = X, <u>Centerline</u> Z, <u>Part Face</u>

Setup Description: Operation No. 2

Material: Brass

Machine the right end including all external work

Tool # and Offset #	Tool Orientation #	Description	Insert Specification	Comments
T0101	3	O.D. Turning Tool	80° Diamond $\frac{1}{32}$ nose radius	500SFM
T0202	3	O.D. Turning Tool	55° Diamond $\frac{1}{32}$ nose radius	600SFM
T0505 T0515	8	O.D. Grooving Tool	$\frac{1}{8}$ wide $\frac{1}{64}$ radius	400SFM
T0707	3	O.D. Threading Tool	60° V-form threading tool	285SFM

Program for the Second Operation

O0027

(ROUGH OD TURNING TOOL)

N1G50S2000

N2T0100M41

N3G96S500M03

N4G00X3.2Z.2T0101M08

N5G01Z.015F.05

N6X-.0624F.014

N7G00X3.2Z.05

N8G71U.03W.005P009Q016D1500F.014

N9G00X1.8161

N10G01Z0

N11X1.9925W-.0882F.004

N12Z-3.8932F.008

N13X2.1105Z-4.0F.004

N14X2.8436F.008

N15X3.0W-.0782F.004

N16X3.2

Part 3 Programming Examples for Lathes

N17G28U0W0T0100M09

N18M01

(FINISH OD TURNING TOOL)

N19G50S2000

N20T0200M41

N21G96S600M03

N22G00X2.2Z.05T0202M08

N23G01Z0.F.05

N24X-.0624F.008

N25G00X3.2Z.05

N26G70P9Q16

N27G28U0W0T0200M09

N28M01

(OD GROOVING TOOL)

N30G50S2000

N31T0500M41

N32G96S400M03

N33G00X2.1Z.2T0505M08

N34Z-2.765

N35G01X1.51F.006

N36G00X2.1

N37Z-2.809

N38G01X1.9925F.008

N39U-.058W.029F.004

N40X1.5F.006

N41W.005

N42G00X2.1

N43Z-2.721T0515

N44G01X1.9925F.008

N45U-.058W-.029F.004

N46X1.5F.006

N47W-.028

N48G00X2.1

N49G28U0W0T0500M09

N50M01

(OD THREADING TOOL)

In lines N33 and N43 the application of two different offsets for both sides of the groove allows a better control of its width.

Note: Values of offsets in the direction of the X axis must be equivalent.

N51T0700M41

N52G97S573M03

N53G00X2.08Z.5T0707M08

N54G76X1.896Z-1.5I0K.0485A60D140F.083333

N55G28U0W0T0700M09

N56T0100

N57M30

Part 3 Programming Examples for Lathes

Example Of Cutting A Three-Start Thread

The following example is given for machining a three-start 1.5-4 Acme thread, as illustrated in the figure below:

Example of Program 28

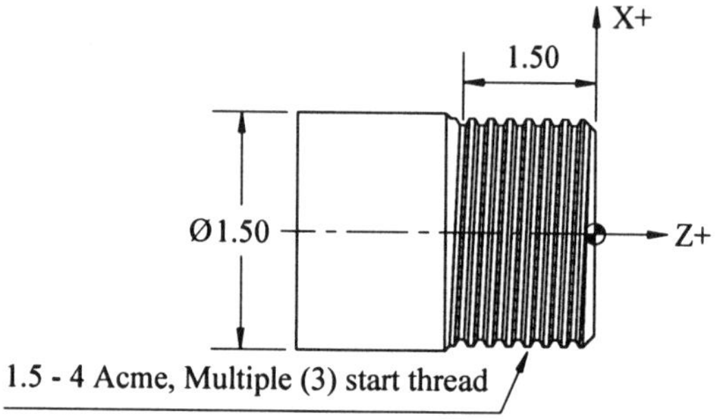

O0028

N1G50S1500

N2T0100M41

N3G97S750M03

N4G00X1.58Z.7T0101

N5M98P1235L3

N6G28U0W0T0100M09

N9M30

Subprogram for program O0028

O1235

N1G00W-.0833

N2G76X1.395Z-1.5I0K.0525A60D140F.250000

N3M99

Figure 138 — 1.5 - 4 Acme, Multiple (3) start thread

Example Of Threading With A Common Tap

Figure 139 — 1/2-13 UNC 2B

Example of Program 29

Machine: Turning Center	Program Number: O0029

Workpiece Zero = X, Centerline Z, Part Face

Setup Description: Operation No. 1

Material: Carbon Steel

Tool # and Offset #	Tool Orientation #	Description	Insert Specification	Comments
T0909		#4 HSS Center Drill		100SFM
T1010		#/# diameter HSS drill		100SFM
T1313		1/2-13 Tap		20SFM

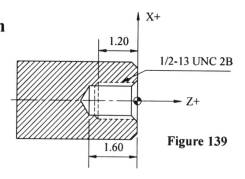

Part 3 Programming Examples for Lathes

Program

O0029

(#4 CENTER DRILL)

N1T0900M41

N2G97S1528M03

N3G00X0.Z.2T0909M08

N4G01Z-.250F.006

N5G00Z1.M09

N6G28U0.W0.T0900

N7M01

(.421 DIAMETER DRILL)

N8T1000M41

N9G97S907M03

N10G00X0.Z.2T1010M08

N11G01Z-1.721F.006

N12G00Z1.0M09

N13G28U0.W0.T1000

N14M01

(1/2-13 TAP)

N15T1300M41

N16G97S76M03

N17G00X0.Z.3T1313M08

N18G32Z-1.4F.0769M05

N19Z.2M04

N20G28U0W0T1300M09

N21M30

Notes:

1. The feed rate override must be set at 100% during the tapping, if function G01 is used.

2. A floating tap holder, which can float and expand forward and backward, should be used for the tapping operation.

Example Illustrating The Application Of The Tool Nose Radius Compensation (G41 and G42)

Functions G41 and G42 refer to the tool nose radius compensation. These functions were described earlier here, in Part III, Application of Tool Nose Radius Compensation.

In short, when using the two initial points in programming, we are limited to programming as if the tool radius is equal to zero. When programming with tool nose radius compensation, the part profile may be programmed. Function G41 refers to the tool radius compensation in a left direction, whereas function G42 refers to the tool radius compensation in a right direction, as shown below. The block format is:

G41X(U) ... Z(W) ...

G42X(U) ... Z(W) ...

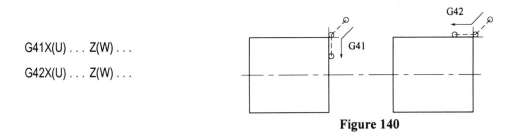

Figure 140

Part 3 Programming Examples for Lathes

The value of the tool nose radius is entered in the offset number to which corresponds with the tool number. After pressing the OFFSET SETTING button on the control, a list of geometry offsets (including columns for R and T) appears on the screen. Under column R, enter the tool nose radius. Under column T, enter the tool orientation number as described earlier in the Application of Tool Nose Radius Compensation section. After completing the turning process, cancel the tool radius compensation with function G40.

Example of Program 30

O0030
N1G50S2000
N2T0100M41
N3G96S500M03
N4G00G41X1.6Z.1T0101
N5G01Z0F.016
N6G00X0F.008
N7G00Z.1
N8G42X1.3
N9X1.5Z-.05
N10Z-.5
N11X2.5
N12G00G40X3.5.Z5.0
N13G28U0W0T0100M9
N14M30

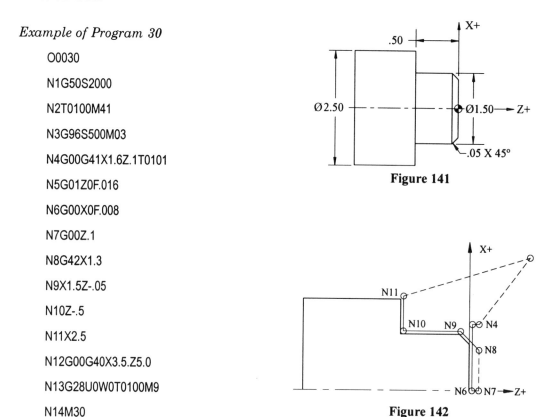

Figure 141

Figure 142

Notes: Functions G41 and G42 are used in turning when tapers and radii must be closely controlled. For all straight turning and facing, initial point programming will obtain the desired results. In examination of the previous program, the only place it is necessary is along the chamfer. Furthermore, G41 applies only to facing while G42 applies for the chamfer and turning. In the following figure, the resultant path is given. Facing to the center of the part along the X axis and then, after withdrawal, perform chamfering and the turning of the remaining part of the shaft. Examine the tool motions with the application of both G41 and G42.

In the above figure, function G41 is activated for facing in line N4 and for turning in G42 is active in lines N8 through N11.

Part III Study Question

1. In NC (tape) controlled machines, the M02 command rewinds the tape to its start. How is a CNC machine commanded to return the program to its start?

   a. M03   b. M08   c. M30   d. M05

2. The term "modal commands" means that once the command is initiated it stays in effect until cancelled or replaced by another command from the same group.

   T or F

3. In programming, tool function is commanded by the four digits that follow the letter address T. (Example T0404). What do these two sets of numbers refer to?

4. Of the following choices, which is the best method for compensating for dimensional inaccuracies caused by tool deflection or wear?

   a. Geometry offsets   c. Tool length offsets

   b. Wear offsets   d. Absolute position register

5. When the rough turning cycle G71 is used, which letters identify the amount of stock to leave for finish for pass X-axis and Z-axis respectively?

   a. U and V   b. X and Z   c. P and Q   d. U and W

6. Sequence (N) numbers in programs may be omitted entirely and the program will execute without any problem.

   T or F

7. The default (at machine start-up) feed rate on lathes is typically measured in

   a. Cutting speed   c. Inches per Minute in/min

   b. Inches per revolution in/rev   d. Constant surface speed

8. The advantage of using G96 Constant Cutting Speed in turning is that as the diameter changes (position of the tool changes in relation to the centerline) the RPM increases or decreases to accomplish the programmed cutting speed.

   T or F

9. The preparatory function G50 relates to two things in programming.

   a. Absolute positioning setting and maximum spindle RPM setting

   b. Absolute position setting and constant spindle RPM

   c. Work offset and tool offsets

   d. Absolute position setting and Constant Surface Speed

10. When incremental programming is required in turning diameters and facing, which letters identify the axis movements respectively?

    a. I and J   b. U and W   c. I and K   d. U and V

Part 3 Study Questions

11. Which M Codes are used to activate and deactivate a subprogram respectively?
a. M03 and M04   b. M2 and M30   c. M98 and M99   d. M41 and 42

12. What M-code is listed in the last line of a subprogram?
a. M30   b. M99   c. M98   d. M02

13. What two letters identify the incremental distance from the starting point to the arcs center in G02/G03 programming?
a. I and J   b. I and K   c. J and K   d. X and R

14. When programming arcs with modern CNC controllers I, J and K can be omitted and replaced by R. If the arc is greater than 180°, what must be added to the R command?

15. Fixed cutting cycle G90 is limited to orthogonal tool movements (no contouring or chamfers are allowed)
T or F

16. When using the Multiple Repetitive Cycle, Rough Cutting Cycle, G71, U and W represent the stock allowance for finishing. What cycle is required to remove the stock allowance?
a. G90   b. G73   c. G76   d. G70

17. What is the major reason for selection of the Pattern Repeating Cycle, G73, as opposed to the Rough Cutting Cycle, G71?
a. It is required to make the finish allowance cut on X and Z
b. This cycle is well suited where an equal amount of material is to be removed from all surfaces.
c. G73 is limited to orthogonal cuts while G71 can cut radius and chamfers
d. G71 is limited to rough cutting only while G73 is required to remove stock allowances

18. When programming arc and angular cuts using tool nose radius compensation, G41 and G42, which is used for facing and turning when tool is mounted above the part centerline.
a. G41 facing, G42 turning
b. G42 facing, G41 turning
c. Neither, the tool nose radius amount must be calculated and programmed to compensate.
d. G41 for facing if contours or angles are involved, G42 for turning if contours or angles are involved.

19. Tool tip orientation needs to be identified in the controller when TNRC is used to program functions G41 or G42. How is this information input?
a. The number is input by R into the program
b. The number is input by T into the offset page
c. The number is input by R into the offset page
d. The number is input by T into the program.

PART 4

Programming Computer Numerically Controlled Machining Centers

Programming Computer Numerically Controlled Machining Centers

In this section, programming of Machining Centers will be presented. In Part I, the steps necessary for programming were identified. In this section, these steps will be followed to create programs. Individual programming words and codes will be defined and demonstrated with examples. This section is written as if the program manuscript is being created manually, line-by-line. This will help you understand the programs and how they are made. In reality though, the most common method for creating programs today is by using CAD/CAM, which is covered in Part V of this text. One of the first things that needs to be dealt with in the programming process is the tool.

Tool Function (T-Word)

The tool function is utilized to prepare and select the appropriate tools from the tool magazine. In order to describe the tool, the address T is followed by one or more digits that refer to the pocket numbers in the tool magazine.

Example:

T05 = tool number 5

Please note that on most modern controllers, it is not necessary to use the leading zero in a tool call; thus, T5 has the same meaning as T05.

Tool Changes

A tool change is specified in the program by the miscellaneous function M06. To initiate a tool change, first call for the desired tool number and then use the miscellaneous function M06 to execute the change.

Example:

N01...T01 (TOOL IN THE READY POSITION)

N02...M06 T02 (ACTUAL TOOL CHANGE) (NEXT TOOL IN THE READY POSITION)

N03...

N04...M06 T03

N05...

Part 4   Programming CNC Machining Centers

In block N01, the requested tool is positioned in the tool magazine to a ready state (waiting position). In block N02, tool T01 is automatically installed into the spindle and, in the mean time, tool T02 is positioned to a ready state in the tool magazine for the next tool change. In block N03, tool 1 performs programmed work. In block N04, a tool change takes place. Tool T02 is installed into the spindle, while tool T01 is returned to the tool magazine and tool T03 is positioned to a ready state in the tool magazine for the next tool change. In block N05, tool T02 performs the programmed work.

Please note that on most modern controllers it is not necessary to use the leading zero in a miscellaneous function call; thus, M6 has the same meaning as M06.

Feed Function (F-Word)

The F-word is utilized to determine the work feed rates. This program word is used to establish feed rate values and precedes a numeric input for the feed amount in inches per minute (in/min) (mm/min) or, inches per revolution (in/rev) (mm/rev). The value that is set by this command stays effective until changed by reentering a new value for the F-word.

Example:

   F20 = a feed rate of 20 inches per minute (in/min)

   F.006 = a feed rate of .006 inches per revolution (in/rev)

If the function G20 (data in inches) is active, such a notation refers to the feed speed of 20 in/min; whereas, with the function G21 (data in millimeters), the notation refers to a feed rate of 20 mm/min.

With rapid traverse G00, the machine traverses at the highest possible feed rate that is specified in control memory (actual rates depend on the design of the machine). In the case of feed rate motion, G01, the value of the feed rate must be accurately specified. The value of the feed rate is determined in one of two ways: inches per minute (in/min) (mm/min) or, inches per revolution (in/rev) (mm/rev) of the spindle. The default feed setting is inches per minute.

Examples:

F20.0 = 20.0 inches of feed per minute

F500. = 500 millimeters of feed per minute

F2.0 = 2.0 inches of feed per minute

F50 = 50 millimeters of feed per minute

F.02 = twenty thousandths inch of feed per revolution

F0.50 = millimeters of feed per minute

F.002 = two thousandths inch of feed per revolution

F0.050 = millimeters of feed per minute

Part 4   Programming CNC Machining Centers

Spindle Speed Function (S-Word)

The letter address S is followed by a specified value in revolutions per minute (rev/min).

Example:

S2100 (specifies 2100 rev/min)

One or more digits following the letter address S are used for the value of the rotational speed. If S0 is input, this command deactivates the spindle rotation and leaves it in a neutral position so that the spindle can be rotated manually, depending on the machine tool. This is quite useful, especially for "dialing-in" the workpiece holder to establish workpiece zero coordinates. The value of S is specified in revolutions per minute (rev/min). Some machines are assigned two ranges of speeds, low or high, others may have three or more ranges. Depending on the given value of the rotational speed, the machine automatically adjusts to the appropriate range, as seen in the following rotational speeds:

In practice, most manufacturers overlap the low and high range, for example:

30-1200 (rev/min) low range

800-4000 (rev/min) high range

Preparatory Functions (G-Codes)

Preparatory functions are the G-codes that identify the type of activities the machine will execute. A program block may contain one or more G-codes.

The letter address G and specific numerical codes allow communication between the controller and the machine tool. This combination of letters and numerical values is commonly called G-Code. In order to perform a specific machining operation, a G-Code must be used. There are two types of G-Codes, modal and non-modal. Modal commands remain in effect, in multiple blocks until they are changed by another command from the same group; whereas, non-modal commands are only in effect for the block in which they are stated.

For example:

Group 00, are non-modal "One-Shot" commands.

Group 01, are modal commands.

There are several different groups of G-codes as indicated in column 2 of Part IV, Chart I. One code from each group may be specified in an individual block. If two codes from the same group are used in the same block, the first will be ignored by the control and the second will be executed. There are G-codes that are active upon startup of the machine indicated by an asterisk (*) in the chart. A safety block is commonly placed in the first line of the program where cancellation codes are used to cancel all G-codes that have been in effect in prior programs. Typically, they are: (G40) cutter compensation can-

Part 4   Programming CNC Machining Centers

cellation, (G49) tool length compensation cancel and (G80) canned cycle cancellation and also (G17) XY plane selection. This cancellation is important because of modal commands that stay in effect until either cancelled or replaced by a command from the same group. It is also a good idea to insert this safety block after tool change blocks, in case of the need to rerun a single operation from within the program. By doing this there is very little chance of modal commands remaining active. The digits following the letter address G identify the action of the command for that block. (Also see the section, Explanation of the Safety Block.)

If the measurement system is changed say from the G20 inch to G21 metric system, then G21 will be in effect at the next startup of the machine.

Chart 1

Code	Group	Function
*G00	01	Rapid traverse positioning
*G01	01	Linear interpolation
G02	01	Circular and helical interpolation CW (clockwise)
G03	01	Circular and helical interpolation CCW (counterclockwise)
G04	00	Dwell
G09	00	Exact stop
*G15	17	Polar coordinates cancellation
G16	17	Polar coordinates system
*G17	02	XY plane selection
G18	02	ZX plane selection
G19	02	YZ plane selection
G20	06	Input in inches
G21	06	Input in millimeters
*G22	04	Stored stroke limit ON
G23	04	Stored stroke limit OFF
G27	00	Reference point return check
G28	00	Reference point return
G29	00	Return from reference point
G30	00	Return to second, third, and fourth reference point
G33	01	Thread cutting
G37	00	Automatic tool length measurement
*G40	07	Cutter compensation cancel
G41	07	Cutter compensation left side
G42	07	Cutter compensation right side
G43	08	Tool length offset compensation positive (+) direction

Part 4 Programming CNC Machining Centers

G44	08	Tool length offset compensation negative (-) direction
G45	00	Tool offset increase
G46	00	Tool offset decrease
G47	00	Tool offset double increase
G48	00	Tool offset double decrease
*G49	08	Tool length offset compensation cancel
*G50	11	Scaling cancel
G51	11	Scaling
G52	00	Local coordinate system
G53	00	Machine coordinate system
*G54	14	Work coordinate system 1
G55	14	Work coordinate system 2
G56	14	Work coordinate system 3
G57	14	Work coordinate system 4
G58	14	Work coordinate system 5
G59	14	Work coordinate system 6
G60	00	Single direction positioning
G63	15	Tapping mode
G68	16	Rotation of Coordinate System
*G69	16	Cancellation of Coordinate System Rotation
G73	09	High Speed Peck drilling cycle
G74	09	Reverse tapping cycle
G76	09	Fine boring cycle
*G80	09	Canned cycle cancel
G81	09	Drilling cycle, spot drilling
G82	09	Drilling cycle, counter boring
G83	09	Deep hole drilling cycle
G84	09	Tapping cycle
G85	09	Reaming cycle
G86	09	Boring cycle
G87	09	Back boring cycle
G88	09	Boring cycle
G89	09	Boring cycle
*G90	03	Absolute programming command

Part 4 Programming Numerically CNC Machining Centers

*G91	03	Incremental programming command
G92	00	Setting for the work coordinate system or maximum spindle RPM
*G94	05	Feed per minute
G95	05	Feed per revolution
G96	13	Constant surface speed control
*G97	13	Constant surface speed control cancel
*G98	10	Canned cycle initial level return
G99	10	Canned cycle initial R-level return

The items marked with an * in the chart above are active upon startup of the machine or when the RESET button has been pressed.

For G00 and G01, G90 and G91 the initial code that is active is determined by a parameter setting. These are typically set at G01 and G90 for startup condition.

Miscellaneous Functions (M Functions)

Code	Function
M00	Program stop
M01	Optional Stop
M02	Program end without rewind
M03	Spindle ON clockwise (CW) rotation
M04	Spindle ON counterclockwise (CCW) rotation
M05	Spindle OFF rotation stop
M06	Tool change
M07	Mist coolant ON
M08	Flood coolant ON
M09	Coolant OFF
M10	Work table rotation locked
M11	Work table rotation unlocked
M13	Spindle ON clockwise and coolant ON, dual command
M14	Spindle ON counterclockwise and coolant ON, dual command
M16	Change of heavy tools
M19	Spindle orientation
M21	Mirror image in the direction of the X axis
M22	Mirror image in the direction of the Y axis
M23	Cancellation of the mirror image
M30	Program end with rewind
M98	Subprogram call
M99	Return to main program from subprogram

Chart 2

Part 4   Machining Center Coordinate Systems

Miscellaneous Functions (M Functions)

Miscellaneous functions or "M-codes", control the working components that activate and deactivate coolant flow, spindle rotation, the direction of the spindle rotation and similar activities.

Programming in Absolute and Incremental Systems

These two coordinate measuring systems are used to determine the values that are input into the programming code for the X, Y and/or Z program words. They can also be used in the same manner for rotary axes A, B and/or C.

Absolute Programming (G90)

In absolute programming, all coordinate values are relative to a fixed origin of the coordinate system. Axis movement in the positive direction does not require inclusion of the sign; while negative movements do require signs.

Incremental Programming (G91)

In incremental systems, every measurement refers to a previously dimensioned position (point-to-point). Incremental dimensions are the distances between two adjacent points.

The coordinate notations for the points on the drawing (in absolute and incremental systems) appear in the chart as follows:

Figure 1

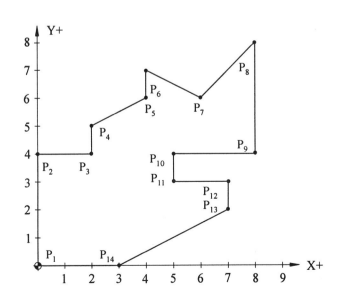

Part 4 Machining Center Coordinate Systems

	For G90		For G91
P_1	G90 X0 Y0	P_1	G91 X0 Y0
P_2	G90 X0 Y4	P_2	G91 X0 Y4
P_3	G90 X2 Y4	P_3	G91 X2 Y0
P_4	G90 X2 Y5	P_4	G91 X0 Y1
P_5	G90 X4 Y6	P_5	G91 X2 Y1
P_6	G90 X4Y7	P_6	G91 X0 Y1
P_7	G90 X6 Y6	P_7	G91 X2 Y-1
P_8	G90 X8Y8	P_8	G91 X2 Y2
P_9	G90 X8 Y4	P_9	G91 X0 Y-4
P_{10}	G90 X5 Y4	P_{10}	G91 X-3 Y0
P_{11}	G90 X5 Y3	P_{11}	G91 X0 Y-1
P_{12}	G90 X7 Y3	P_{12}	G91 X2 Y0
P_{13}	G90 X7 Y2	P_{13}	G91 X0 Y-1
P_{14}	G90 X3 Y0	P_{14}	G91 X4Y-2
P_1	G90 X0 Y0	P_1	G91 X-3 Y0

Chart 3

Programming of Absolute Zero Point (G92)

Workpiece zero is an optionally chosen position for the origin of the coordinate system on the drawing and is determined by the programmer. Workpiece zero, in most cases, is selected as the most suitable point on a given workpiece. For example, it may be in the center, as in Figure 2 (if the part is symmetrical), or, at intersecting edges as in Figure 3.

Example:

Figure 2

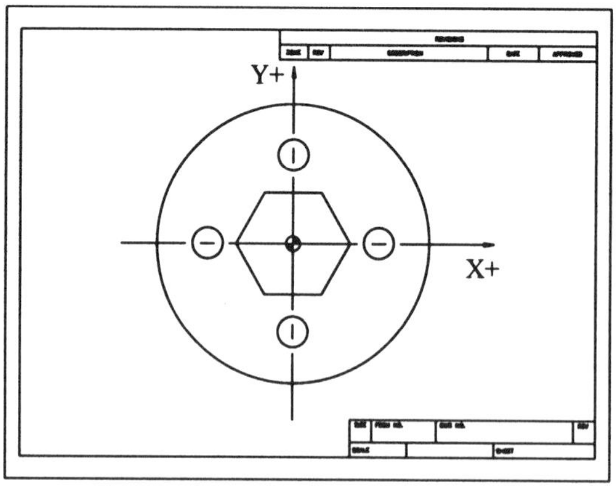

All tool motions are performed with respect to the workpiece zero. The G92 format is:

G92 X...Y...Z...

The coordinate values input to the X, Y and Z program words assigned in the block with function G92, correspond with the distance from workpiece zero to machine zero for each axis.

Part 4 Machining Center Coordinate Systems

Setting G92

A way to determine these values, is to home the machine for each axis and then press the "POSITION" function button on the control panel and a position readout appears on the screen. Then, press the "PAGE" button to find the corresponding absolute position readout. Now, check to see whether all of the values shown on the readout are equal to zero. If not, then press button "X" followed by the soft key labeled "ORIGIN". Follow the same pattern for the remaining axes Y and Z.

The steps for establishing the workpiece zero of the coordinate system for the first drawing are as follows:

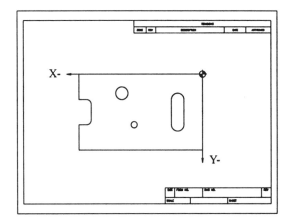

Figure 3

Attach a dial indicator to the spindle of the machine. Then, switch to the handle mode (MPG as described in the Operator section) to move the position until the axes are centered over the hole and then dial-in the hole.

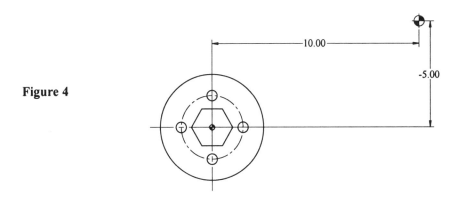

Figure 4

Then, assign the values of X and Y that are shown on the readout at the beginning of the program to the function G92 X10 Y5. The values of X and Y correspond to the distance between machine zero and the workpiece zero. Workpiece zero in the example illustrated on the second drawing is found in a similar manner. Position to machine zero with respect to all axes. Then, install an edge finder to the spindle of the machine.

Set the RPM at 1500 and move the spindle along the path shown on the drawing by the dashed line. As you approach the edge, reduce the speed of the traverse in order to

Part 4   Machining Center Coordinate Systems

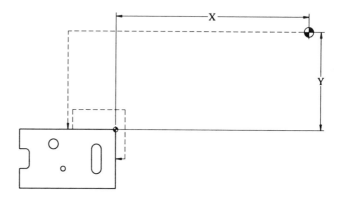

Figure 5

accurately spot the deflection of the edge finder. As soon as the first deflection of the edge finder is noticed, interrupt the feed along the Y axis and read the value of Y from the position readout. To this value, add the radius of the tip of the edge finder, and then enter this entire value under the Y assigned to the function G92.

The same procedure is applied for the edge positioned along the X axis. As soon as the edge finder deflection is noticed, read this value from the position readout. Add to that the value of the radius of the tip of the edge finder and enter it under the X axis assigned to the function G92. The values of X and Y so obtained correspond to the distance along the X and Y axes between machine zero and the workpiece zero. The value of Z assigned to function G92 is usually zero, which means that the workpiece zero corresponding to the Z axis is in the position of machine zero for the same axis. The values entered into the Tool Length Offsets compensate for the movements of the Z axis.

The values obtained for X, Y, and Z, assigned to function G92, are always entered at the beginning of the program and it is also advisable to enter it after each tool change, so if the program is restarted at the tool change then the coordinate system will be reestablished.

G92 X...Y...Z...

Function G92 is best used with less complicated programs in which only one workpiece zero is utilized. If the contour of the machined workpiece is complicated and difficult to program, it is best to use G54 through G59 for workpiece zeros. This is the most widely used method for assigning workpiece zeros today.

In order to specify multiple workpiece zeros, use the following functions:

G54	first workpiece zero
G55	second workpiece zero
G56	third workpiece zero
G57	fourth workpiece zero
G58	fifth workpiece zero
G59	sixth workpiece zero

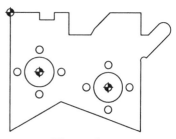

Figure 6

Part 4   Machining Center Program Creation

Functions G54, G55, G56, G57, G58, and G59 are described in detail later in this chapter and for the process of setting G54 – G59 refer to Part II Operation.

Program Creation

Program Number

Each program is assigned a number. The capital letter O is reserved for the program number identification and is usually followed by four digits, which specify the actual program number. For example, to create a program with the number 1234, the programmer must input the letter address O, and then the number 1234 (O1234). All programs require this format.

Examples:

O0001 = program number 1

O0014 = program number 14

A common mistake made here, is to enter zero (0) instead of the letter O, which results in an alarm on the control system.

Comments

Comments, to aid the operator, may be added to the program by using the parenthesis. The comments or data inside the parenthesis will not affect the execution of the program in any way. A very common place to add comments is at the program number, to identify a part number, or at a program stop direct the operator in some way. An End-of-Block character (EOB) is typically required after the parenthesis.

Examples:

O0001

(PN587985-B)

N9M00

(REMOVE CLAMPS FROM OUTSIDE OF PART)

Block Number

The letter "N" is reserved to identify the program line sequence numbers (block numbers) and precedes any other data in a program line. For each line in a program, a block number is assigned sequentially. For example, the first block of a program is labeled N1, the second N2 etc. Typically the program block numbering system is sequenced by an increment of other than one. An example of this is sequencing by five, where line one is labeled N5 and line two is labeled N10. The intent of an increment of other than one is to allow for adding additional blocks of data between the increments. Doing this is sometimes advantageous when editing programs. It is important to note that block numbering

Part 4 Machining Center Program Creation

is not necessary for the program to be executed. The program blocks, even if not numbered, will be executed sequentially during automatic cycle of the machine. Removing block numbers is sometimes helpful when a program is too large to fit into the controller resident memory, as each character of a program takes space and large programs have block numbers into five digits. It is also worth noting that it is a common practice to place block numbers at a tool change command. This enables the operator to restart the execution of the machining program at a specific tool change where, otherwise, the entire program would have to be executed in order to rerun a specific operation and time would be wasted. Block numbers enable movement within the program to enter offsets, verify data, or search for a block in the program. They are often referenced by the control in the case of programming error enabling the programmer to search to the problem directly by block number. Block numbers are not required at all for the program to work but they are necessary for restarting the program at a specific place.

Subprogram

If a certain part of the program can be used repeatedly within the program, assign a number for this part and list it separately, calling for it whenever it's needed. This part of the program is called a subprogram. Subprograms greatly simplify programming and decrease the amount of data that must be placed into the controller memory. A subprogram is called up from control memory by the function M98 and the letter P, both of which are entered in the main program. The letter P refers to the program number of the subroutine call. Similarly, as in the main program, in order to enter the subprogram number, you must first write the letter O and then the number. For each subprogram number, the four spaces after the letter O are reserved just as with the main program. Ordinarily, there is no workpiece coordinate system established within the subprogram because of its dependence on the main program.

Example:

O0012

N0001

. . .

. . .

N0126M99

Function M99 refers to the end of the subprogram and returns to the main program block following the block containing the subprogram call in the main program.

Example:
M98P0012

This is a call for subprogram No. 12.

Part 4 Machining Center Program Creation

Main program	Subprogram	Subprogram
O0001		
N0001G54 X..Y...		
N0002 ...		
N0003 ...		
...		
...		
...		
...	O0012	O0013
...	N0001 ...	N0001 ...
...	N0002 ...	...
N0125M98P0012	...	...
N0126 ...	...	...
...	N0018M98P0013	...
...	N0019 ...	...
...	...	...
N0287M30	N0075M99	N0032M99

In the above example, enter the subprogram call in the program block N0125 of the main program, and then use function M98 to call up subprogram O0012. Work is then performed according to the commands in the subprogram until block N0018 is reached. The processing is then transferred to the next subprogram O0013.

After the execution of subprogram O0013, the program is returned to block N0019 of the subprogram O0012, which executes the remaining information blocks of this subprogram. From block N0075 of this subprogram, the program is returned to block N0126 of the main program, which then executes the remaining part of the main program. As many as four levels of subprograms can be linked together and is called nesting. To repeat a given subprogram twice, enter the following:

M98P0012L2

L2 repeat subprogram twice

For the controller identified in this edition (16M and 18M) the syntax is slightly different. The subprogram call works like this:

M98P00010012

Where M98 is the call for the subprogram, P0001 is the number of repeats and 0012 is the subprogram number. It is highly recommended that the programmer study the manufacturer manuals for specific instructions on using subprograms.

Part 4   Machining Center Program Creation

Program End

The difference between M02 and M30 is that M02 refers to the program end, while M30 refers to the program end and a simultaneous return to the program head. Both commands are found in the final line of the main program only. Note: on some controls, the M02 behaves the same as M30 for compatibility with older programs.

The Link Between Functions G92 and G43

As a rule, workpiece zero for the Z axis is placed within the machining envelope in relationship to machine zero. Workpiece zero for a specific program for the Z axis is, as a rule, on the top most surface of the machined part. At first glance, a contradiction between these two statements seems to exist. However, both statements are true, because there is a direct relation between the functions G92 and G43 if we refer them both to the Z axis. Workpiece zero for the Z axis is to be in the same position as it is for machine zero for that axis, this is applied to all the tools. Due to different tool lengths, in order to transfer program zero along the Z axis from machine zero to the surface of the workpiece, you must apply function G43. Function G43 is a tool length function (tool length offset). Offset number H . . . (H01, H02) is always assigned to function G43. In order to simplify program execution, offset number H, as a rule, is to be the same as the tool number for each corresponding tool. The value of offset H is entered into the offset registers in the computer memory (for example, H01 = -11.1283). The value of the tool length offset for a given tool corresponds to the distance between the tool tip and the surface of the workpiece.

In order to determine the value of offset H for a specific tool, zero the machine with respect to the Z axis with the tool in the spindle to be measured. Zero the position readout for Z; then, manually move the tool along the Z axis to the surface of the workpiece in such a way that the tool tip touches the surface of the workpiece. This distance is what determines the value of offset H for a given tool, and it is registered in the offset table with the number corresponding to the offset number in the program.

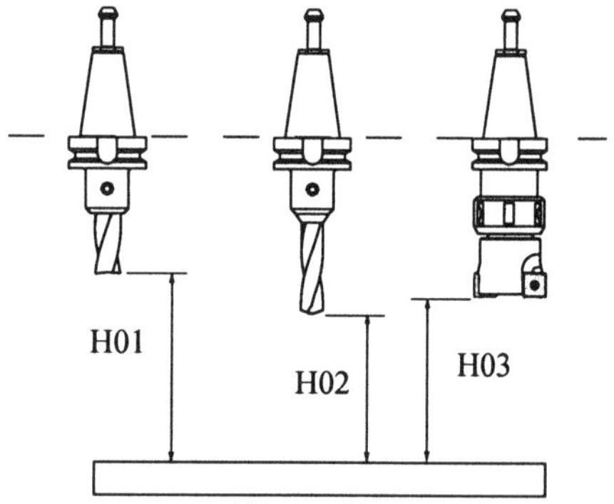

Figure 7

Function G43, with the assigned offset number H, must be entered in the program before the tool does any work. If the offset number H . . . is not entered, the tool will perform the work with the previous tool length offset. If the value of the tool length offset for the previous tool is smaller than the value of the tool length offset of the working tool, then the tool will not approach the material.

CAUTION: If, however, this value is greater than the value of the tool length offset of the working tool, the tool will then rapidly advance toward the workpiece and crash into it, causing damage to the tool, the workpiece, and the holding equipment.

Part 4 Machining Center Program Creation

Example of Program No. 1

Two holes, with diameters of .500 inch each, are to be drilled in a steel plate having the dimensions 9 x 4 in. The position of the holes can be seen in the drawing below.

YOUR COMPANY
CNC SETUP SHEET

Date:	Prepared By:
Part Name:	Part Number
Machine:	Program Number:

Workpiece Zero X __________ Y__________ Z__________

Setup Description:

Tool #	Description	Offset	Comments

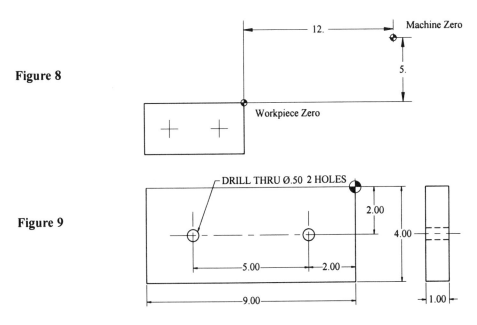

Please note the use of the CNC Setup sheet above as described in Part I of this book. For the remainder of this section it will be used for all of the examples. The top portion including the Title, Date, Prepared By, Part Name and Part Number will be omitted and only identification for each tool will be given to save space.

Figure 8

Figure 9

Part 4   Machining Center Program Creation

Program

O0001

N1G90G20G80G40G49

N2G92X12.Y5.Z0T02S900M03

N3G00X-2.Y-2.0

N4G43Z1.H01M08;

N5G81G99Z-.219R.1F4.6

N6G00X-7.Y-2.0

N7G80Z1.M09

N8G28Z1.0

N9M01

N10M06

N11G92X12.Y5.Z0

N12G00G90X-2.Y-2.S620M03

N13G43Z1.H02M08

N14G81G99Z-1.25R.1F7.0

N15G00X-7.Y-2.0

N16G80Z1.M09

N17G91G28Z0M05

N18G28X0Y0

N19T01M06

N20M30

Tool one, the number 3 center drill, is mounted in the spindle prior to beginning.

In block N1, functions G90, G20, G80, G40 and G49 are defined as follows.

G90; establishes absolute dimensioning for the coordinate system.

G20; establishes the inch system of measurement.

G80; cancels all canned cycle functions and is entered at the beginning of the program to ensure that all cyclic functions from the previous program are no longer in effect.

G40; cancels cutter diameter compensation functions, G41 and G42 are cancelled.

G49; cancels the tool length compensation functions, G43 and G44 are canceled.

In block N2, distances between machine zero and workpiece zero (X12., Y5.) are entered into the controller memory. In addition, the spindle speed and direction of the rotation (S900, M03) are defined, while the program word T02, positions tool two in the

Part 4   Machining Center Program Creation

magazine for replacement in the spindle (the tool is changed in block N10). Block N3 directs the tool to the proper position for drilling the first hole.

In block N4, the values of tool length offset are read (center drill), while rapid traverse of the spindle along the Z axis (with a value of Z = 1.000) is performed to a position above the workpiece. The miscellaneous function, M08, activates the flood coolant flow. In block N5, a hole is drilled with a drill feed equivalent to 4.6 IPM. In block N6, a second hole is drilled. Block N7 cancels canned cycle function G81, raises the tool to the initial reference plane position of Z1.0, and deactivates the coolant flow with M09. In line N8 the machine returns to zero with respect to the Z axis.

In block N9, the programmed machine work is stopped by function M01, but only if the "optional stop" button on the operator control panel is on. The purpose of this block is primarily to confirm whether tool T01 has performed the work that has been assigned in the program. This block usually appears before the tool change. This stoppage of the program execution gives the operator an opportunity to perform in-process inspection of the completed operation. The reason this block appears before the tool change in the program, is to give a clear picture of the programmed movements on the part geometry and allows inspection of the workpiece by measuring to see where the work of each particular tool is performed.

In block N10, tool 1 is replaced by tool 2 in the spindle. Up to now, only one program segment for one tool has been executed. Now, by comparing the remaining part of the program, notice that it contains many similar elements for the second tool, the .500 diameter drill.

The program segment consists of a few characteristic elements that play the following roles:

- Establishment of the absolute or incremental coordinates system, spindle RPM and rotation direction.
- Executes the tool length offset given and activates the flow of the coolant.
- Determines the drilling canned cycle.
- Determines the tool's work path to consecutive holes in the pattern to be drilled.
- Positioning of the tool to the consecutive locations.
- Cancels any canned cycles and deactivation of the coolant flow.
- Command of the X and Y axes to return to machine zero position.
- Tool change.

Part 4 Machining Center Program Creation

Example of Program No. 2

Machine: Machining Center	Program Number: O0002

Workpiece Zero X = <u>Upper Right Corner</u> Y = <u>Upper Right Corner</u> Z = <u>Top Surface</u>

Setup Description:

The workpiece material is aluminum.

Dimensions of the plate before machining are 3. × 3. × 1.5

The workpiece is mounted in the vice on parallels with the stop on the right side.

Tool 1 is placed in the spindle before the work from the program begins.

Tool #	Description	Offset	Comments
T01	Two flute $\frac{1}{8}$ in (.625) diameter High Speed Steel End Mill		SFM = 400
T02	No. 6 Center Drill		SFM = 400
T03	$\frac{1}{2}$ in Drill		SFM = 400

Program

O0002

N10G90G20G80G40G49

N15G92X21.025Y6.127Z0

N20G00X.4Y-.1875S2445M03T02

N25G43Z1.M08H01

N30G01Z-.25F20.0

N35X-2.8125

N40Y-2.8125

N45X-.1875

N50Y.4

N55G28Z1.0.M09

N60M01

N65M06

N70G90G80G40G49

N71G92X21.025Y6.127Z0

N75G00X-1.5Y-1.5S3057M03T03

N80G43Z1.H02M08

N85G81G98Z-.438R.1F12.0

N90G80M09

N95G28Z1.0

N100M01

N105M06

N110G90G80G40G49

N111G92X21.025Y6.127Z0

N115G00X-1.5Y-1.5S3057M03T01

N120G43Z1.H03M08

N125G81G98Z-1.664   R.1F18.0

N130G80Z1.0M09

N135G91G28Z0M05

N140G28X0Y0

N145M06

N150M30

Part 4   Machining Center Program Creation

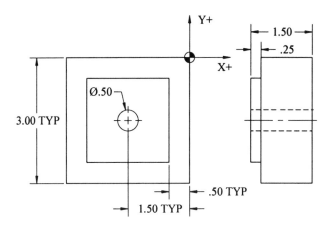

Figure 10

EXPLANATION OF THE SAFETY BLOCK

Block N10 is called the safety block; the name refers to its role in the program. This safety block consists of G-codes that establish the type of coordinate system used (G90 Absolute or G91 Incremental), establishes the measurement system used (G20 inch or G21 metric), cancellation of all canned cycles (G80), cancellation of cutter diameter compensation (G40), and cancellation of tool length compensation (G49). In order to more accurately define the canceling function of G80, study the example program. In block N85, the canned drilling cycle G81 is activated. If the canned drilling cycle is interrupted for some reason (the drill or the workpiece become damaged), the machining must be stopped, in order to exchange the tool and or the workpiece. Then by pressing the "reset" button the program is returned to its head. However, because of the above occurrence, canceling function G80 (included in block N90) has not occurred. (On many machines, RESET will cancel canned cycles.) If machining is started from the beginning of the program, tool one (an end mill of .625 inch) will need to be returned to the spindle and the new drill returned to the tool magazine prior to starting. In the above-described procedure, if function G80 is omitted from block N10, then function G81 is still valid. Thus, all position changes of the end mill (along the X and Y axes) will be executed as if the canned drilling cycle including the assigned parameters of Z and R from block N85, are still valid.

To avoid such mistakes, use the safety block including the canceling functions in the first block of the program and immediately following all tool changes.

A similar situation in which tool radius compensation was applied is described as follows. In programming the milling process, you may establish the movement of the tool by determining the movement of the center of the tool. This approach applies to standard programming. However, you may also program the actual contour of the workpiece. From that, the computer will calculate the path of the center of the mill (or other tool) with a given radius. This means that the actual machine movements will differ from the

Part 4 Machining Center Preparatory Functions

programmed movements by the value of the offset, which is the radius of the tool. Such an arrangement can be applied by use of functions G41 (cutter compensation to the left) and G42 (cutter compensation to the right). If, for any reason, the milling process with the assigned tool radius compensation is interrupted, then the next tool used, for example, a drill, will be positioned with a displacement value equal to the mill radius compensation previously used. In order to avoid such situations, employ the cancellation function G40 in the safety block.

The same scenario can be applied to tool length compensation values that are activated in the program by the letter address H followed by the number of the tool as shown in block N25. If this offset value is not cancelled by G49 in the safety block, it too will remain valid. Note: If function G28 is used prior to tool change as in line N55, the tool length offset is cancelled.

To avoid such mistakes, use the safety block, including the canceling functions, in the first block of the program and immediately following all tool changes.

Overview of Preparatory Functions (G Functions)

These preparatory functions, often called G-codes, are a major part of the programming puzzle. They identify to the controller, what type of machining activity is needed. For example, if a hole needs to be drilled, function G81 may be used, or if programming in the incremental coordinate system is required, then function G91 is used. These codes, along with other data, control motion.

The motion of the axes of a machine may be performed along a straight line, an arc or a circle.

Codes G00 and G01 allow axes movement along a straight line.

Codes G02 and G03 allow axes movement along a circular path of motion.

Rapid Traverse Positioning (G00)

The rapid traverse function is entered to relocate the tool from position A to position B along a straight line at the fastest possible traverse. The shortest axis movement distance will be accomplished first and, therefore, you must concentrate on the workpiece holding the equipment in order to avoid any collision between the tool and the holding equipment. The path traveled by the tool for G00, X30, and Y20, is as follows:

Figure 11

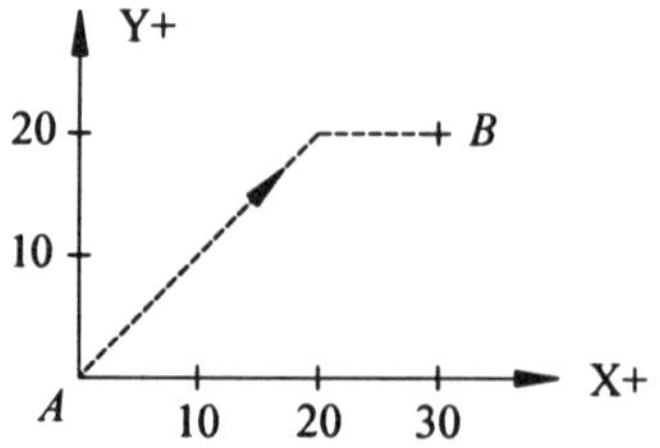

Part 4 Machining Center Preparatory Functions

If the tool path is uncertain, then position the "RAPID TRAVERSE OVERRIDE" switch at 25 or 50%. This reduces the traverse speed and increases the time allowance for a possible manual interruption of the motion. Programmers should position the Z axis to an acceptable clearance plane of 1.0 inch or any amount necessary to move over clamps or obstructions to eliminate this problem.

By adjusting a system parameter, the movement described above can be changed to a simultaneous or diagonal move of both axes. Consult the manufacturer manual for specific instructions.

Linear Interpolation (G01)

The work function G01 is used to move the tool from point P_1 to point P_2 along one or all of the axes simultaneously, along a straight line of motion and at a given feed rate specified by the F-word.

Example:

G01X10.Y20.F8.0

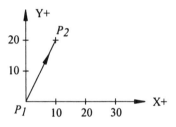

Figure 12

Linear interpolation may be performed along three axes simultaneously. The control system calculates the particular speeds for each axis so that the resulting speed is equivalent to the programmed feed rate.

Example of Program No. 3

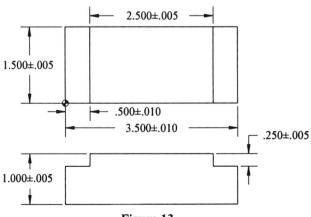

Figure 13

Part 4 Machining Center Preparatory Functions

Machine: Machining Center	Program Number: O0003

Workpiece Zero X = <u>Lower Left Corner</u> Y = <u>Lower Left Corner</u> Z = <u>Top Surface</u>

Setup Description:

Material used is Aluminum 6061-T6, with plate dimensions of $3.5 \times 1.5 \times 1.0$.

$$S = \frac{12 \times SFM}{\pi \times d} = \frac{12 \times 400}{3.14 \times .625} = 2445 \text{ rev/min}$$

$$F = 4 \times .003 \times 2445 = 29 IPM$$

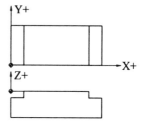

Figure 14

Tool #	Description	Offset	Comments
T01	Two Flute .625 Diameter HSS End Mill		SFM = 400 ft/min Feed =.003 in per flute

Program

O0003

N10G90G20G80G40G49

N15G92X11.0Y6.0Z0

N20G00X.1875Y-.4S2445M03

N25G43Z1.0H01M08

N30G01Z-.250F50.0

N35Y1.9F29.0

N40G00Z1.0

N45X3.3125

N50G01Z-.250F50.0

N55Y-.4F29.0

N60G28Z1.0M09

N65G91G28X0Y0M05

N70M30

Circular Interpolation (G02, G03)

Circular interpolation allows tool movements to be programmed to move along the arc of a circle. When applying the circular interpolation, the plane in which the arc is positioned must be determined initially. To do this, employ preparatory function G17,

Part 4   Machining Center Preparatory Functions

G18, or G19. Then, depending on the direction of the machining, select function G02 to make a clockwise movement along the circle and function G03 to make a counterclockwise movement along the circle. In order to describe the movements of the tool along the circle, apply the following two methods:

1. Determine radius R and the values of the start point coordinates in the given plane.
2. Determine the value of the endpoint of the arc and the values of the incremental distance to the arc center in a given plane.

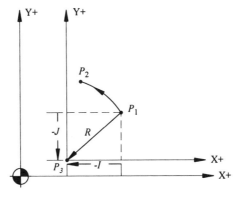

Figure 15

P_1 = the start of an arc

P_2 = the end of an arc

P_3 = the arcs center

R = radius vector

I = the incremental distance to the arc center along the X axis.

J = the incremental distance to the arc center along the Y axis.

The incremental distance is defined as a radius projection onto a given axis. The radius incremental distance is always attached. It begins at the starting point and ends at the center of the circle. It is always directed toward the center of the circle.

Vector projections onto the X axis are identified by use of the letter address I.

Vector projections onto the Y axis are identified by use of the letter address J.

Vector projections onto the Z axis are identified by use of the letter address K.

In Figure 16 an example a vector projection is illustrated. The radius is positioned at the origin of the coordinate system, (see following page).

Note in the example above that the signs (+, -) of the incremental distances I, J, and K depend on the position of the starting point of the arc with respect to the center of the arc, that is, with respect to the coordinate system. If the direction of the vector is consistent with the direction of the assigned axis of the coordinate system, apply the positive sign, if not, apply the negative sign.

Most modern controllers do not require the use of the positive sign. If no sign is present the value is considered to be positive.

Part 4 Machining Center Preparatory Functions

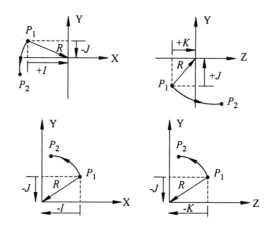

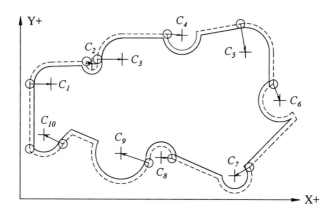

Figure 16

Figure 17

In the Figure 17 above, the tool path begins in the lower left corner of the part. Each arc is identified C_1 through C_{10} with the center point of the arc shown. The arrows indicate direction vector to the arc center. Following are the arc cutting directions and the I and J values:

C_1 = G02, I+, J0
C_2 = G03, I+, J-
C_3 = G02, I+, J0
C_4 = G03, I+, J-
C_5 = G02, I+, J-
C_6 = G03, I+, J-
C_7 = G02, I-, J-
C_8 = G03, I-, J+
C_9 = G02, I-, J+
C_{10} = G02, I-, J+

Part 4 Machining Center Preparatory Functions

Notes: When circular motion is described with radius function R, no sign is required if the arc is less than or equal to 180° (the system defaults to positive unless otherwise specified) and assign a negative value to R (-) if the arc is greater than 180°. The maximum rotation of an arc using R is 359.9°.

A full circle may be accomplished using R by linking two 180° arcs.

If a full circle of 360° is to be performed, then it is necessary to employ the incremental distances of the arc center points for I, J, and K, not radius R, in the program.

Example:

Do not use I, J, or K with R in the same block, because if either I, J, or K are used, they will be ignored by the control, and the tool will follow the arc with the assigned radius of R.

If the value of an entered radius R is zero, an alarm will result.

Cutter radius compensation may be used for circular interpolation. However, it must be called up in a G00 or G01 block preceding the G02/G03 information..

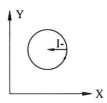

Figure 18

Example of Program No. 4

An example illustrating the application of circular interpolation:

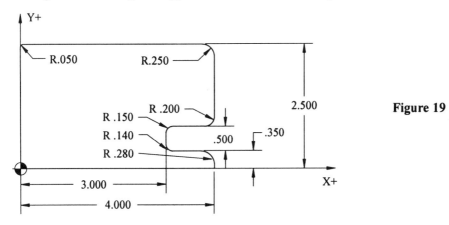

Figure 19

Machine: Machining Center	Program Number: O0004

Workpiece Zero X = <u>Lower Left Corner</u> Y = <u>Lower Left Corner</u> Z = <u>Top Surface</u>

Setup Description:

Material used is Steel and is .150 thick.

The program is limited to a finishing pass.

The workpiece is mounted to a machining fixture by the existing .250 diameter holes not shown in the above print.

F = 4 × .002 × 1375 = 11 in/min

Tool #	Description	Offset	Comments
T02	Four Flute .250 Diameter HSS End Mill		SFM = 90 ft/min Feed =.002 inches per flute

Part 4 Machining Center Preparatory Functions

Program

O0004

N10G90G20G80G40

N15G92X10.0Y7.0Z0

N20G90G00X-.125Y-.2S1375M03

N25G43Z1.0H02M08

N30G01Z-.16F50.0

N35Y2.45F11.0

N40G02X.05Y2.625I.175J0

N45G01X3.75

N50G02X4.125Y2.25I0.J-.375

N55G01Y1.05

N60G02X3.8Y.725I-.325J0

N65G01X3.15

N70G03X3.125Y.7I0.J-.025

N75G01Y.49

N80G03X3.140Y.475I.015J0

N85G01X3.72

N90G02X4.125Y.07I0.J-.405

N95G01Y-.125

N100X-.125

N105G28Z1.M09

N110G91G28X0Y0M05

N115M30

The following is the same program but utilizing radius R:

O0004

N10G90G20G80G40

N15G92X10.0Y7.0Z0

N20G90G00X-.125Y-.2 S1375M03

N25G43Z1.0H02M08

N30G01Z-.1F50.0

N35Y2.45F11.0

Part 4 Machining Center Preparatory Functions

N40G02X.05Y2.625R.175

N45G01X3.75

N50G02X4.125Y2.25R.375

N55G01Y1.05

N60G02X3.8Y.725R.325

N65G01X3.15

N70G03X3.125Y.7R.025

N75G01Y.49

N80G03X3.14Y.475R.015

N85G01X3.72

N90G02X4.125Y.07R.0405

N95G01Y-.125

N100X-.125

N105G28Z1.M09

N110G91G28X0Y0M05

N22M30

Helical Interpolation Using G02 or G03

Helical interpolation allows movements of the tool to be programmed so that it travels along a circular path in the XY plane, with a simultaneously straight-line motion along the Z axis. Such a combination of tool movements with respect to the three axes creates a helix contour and may be used to machine threads. For all practical purposes, this function is limited to machining threads with large diameters because of the required tool diameter.

Command:

G02X . . . Y . . . I . . . J . . . Z . . . F . . . (R . . . may be used in place of I and J)

G03X . . . Y . . . I . . . J . . . Z . . . F . . . (R . . . may be used in place of I and J)

Where

    X = arc ending point

    Y = arc ending point

    I = incremental X axis distance of the arc center from the start point

    J = incremental Y axis distance of the arc center from the start point

    Z = arc ending point

    R = radius of the arc

    F = feed rate

Part 4 Machining Center Preparatory Functions

Function G02 refers to clockwise motion (CW); whereas, function G03 refers to counterclockwise motion (CCW).

Notes on Helical Interpolation

This function may only be applied under the following conditions:

The minor diameter for internal threads or major diameter for external threads, must be machined in a prior operation.

The XY plane is in a circular interpolation.

The Z axis is in a linear interpolation.

The tool must be positioned in the Z axis prior to the helical interpolation.

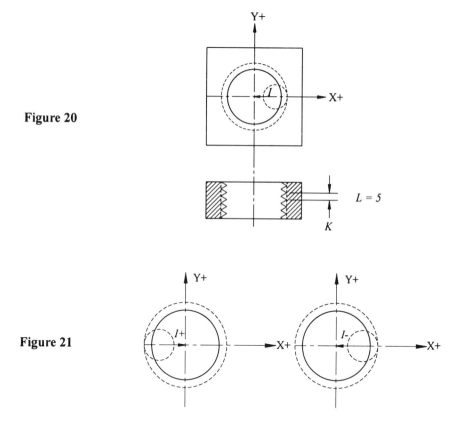

Figure 20

Figure 21

The feed rate designated by the F-word is the circular interpolation for the XY plane.

Climb milling is the preferred machining method.

When one pass of the tool does not complete the full thread, adjustment of the Z axis coordinates of the starting or ending point (depending on machining starting from the bottom or top of the thread), for another pass is required.

If cutter radius compensation is used in the program, G41 or G42 may not be called in the same line as the G02 or G03. For an internal thread, it is typically called in a linear approach move from the center of the diameter moving outward and cancelled at the

Part 4 Programming Numerically CMC

end of interpolation moving back towards the center. This move must be equal to, or greater than the radius of the tool used.

Figure 22 is an example of a tool that may be utilized during threading with the application of helical interpolation.

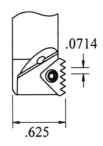

Figure 22

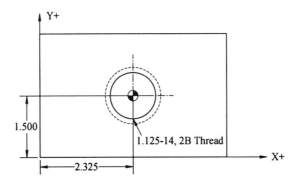

Figure 23

Example of Program No. 5

An example for helical interpolation follows:

Machine: Machining Center	Program Number: O0005

Workpiece Zero X = <u>Part Centerline</u> Y = <u>Part Centerline</u> Z = <u>Top Surface</u>

Material used is .50 thick CD 12L14 steel plate. $S = \dfrac{12 \times V}{\pi \times d} = \dfrac{12 \times 70}{3.14 \times .625} = 428$

The minor diameter was machined in a previous operation and only the threading cycle is needed.

Tool #	Description	Offset	Comments
T01	Special threading tool with a .0714 lead. See Figure above.		SFM = 70 ft/min Feed = 1.5 IPM

Part 4 Machining Center Preparatory Functions

Program

O0005

N10G90G20G80G40G49

N15G92X10.0Y7.0Z0

N20G00X0Y0S428M03

N25G43Z1.H01M08

N30G01Z-.575F50.0

N35G01X.25F3.5

N40G03X.25Y0I-.25Z-.5036F1.5

N45I-.25Z-.4322

N50G01X0F50.0

N55G00Z1.0M09

N60G91G28Z0M05

N65G28X0Y0

N70M30

Note: The value X.25 in block N35 was calculated in the following manner:

The difference between the Z value on line N40, -.5036 and on line N45, -.4322, is equal to the lead of the thread, .0714.

$$X = \frac{1.125}{2} - \frac{.625}{2} = .250$$

Dwell (G04)

Dwell is determined by the preparatory function, G04 and by using the letter address P or X, which corresponds to the time duration of dwell.

Command:

  G04P . . .

  G04X . . .

Example:

  G04X1.   or   G04P1000

The length of time for the dwell is stated by X or P and, in this example, is equivalent to 1 second. P address is programmed in milliseconds (ms) 1 second = 1000 ms.

When the letter address P is used to determine time, do not use the decimal point. Drilling or boring holes, with a specifically defined depth, are two examples in which to apply function G04. Time P and X, in these cases, determine a time period in which the drill, having full rotational speed, stops at the bottom of the hole and accurately removes the excess material from the bottom of the hole.

Part 4 Machining Center Preparatory Functions

Example of Program No. 6

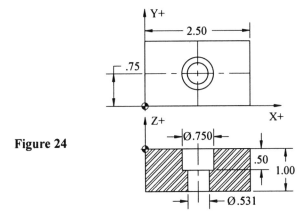

Figure 24

Machine: Machining Center	Program Number: O0006

Workpiece Zero X = <u>Lower Left Corner</u> Y = <u>Lower Left Corner</u> Z = <u>Top Surface</u>

Setup Description:

The material used is 4140 steel.

The part is clamped in a vise.

Tool one is mounted in the spindle prior to beginning.

Tool #	Description	Offset	Comments
T01	No. 4 HSS center drill.		SFM = 75 ft/min Feed = .001 inch per flute
T02	$\frac{17}{32}$ HSS drill		SFM = 75 ft/min Feed = .002 inch per flute
T03	.750 diameter 2 flute HSS end mill		SFM = 75 ft/min Feed = .002 inch per flute

Program

O0006

N10G90G20G80G40G49

N15G92X10.0Y7.0Z0

N20T1M06

N25G00X1.25Y.750S960M03T02

N30G43Z1.H01M08

N35G00Z.1

N40G01Z-.269F1.9

N45G91G28Z0M09

Part 4   Machining Center Preparatory Functions

N50M01

N55M06

N60G90G80G40G49

N65G00X1.25Y.750S564M03T03

N70G43Z1.H02M08

N75Z.1

N80G01Z-1.35F2.25

N85G91G28Z0M09

N90M01

N95M06

N100G90G80G40G49

N105G00X1.25Y.750S400M03T01

N110G43Z1.H03M08

N115Z.1

N120G01Z-.5F3.2

N125G04P300

N130G00Z1.0M09

N135G91G28Z0M05

N140G28X0Y0

N145M30

In block N125, a dwell of .3 seconds is utilized to smooth out the bottom.

Exact Stop (G09)

When this function is used, the machine responds by decelerating to a feed of zero in the axis commanded before executing the next line, thus, checking the programmed end point. This function is used only if you want to obtain a sharp edge around corners in the cutting feed mode. Also, it refers only to the block to which it was assigned.

Polar Coordinate Cancellation (G15)

This command is used to cancel use of the polar coordinate system called by function G16. This command must be programmed in a line, by itself.

Part 4   Machining Center Preparatory Functions

Polar Coordinate System (G16)

The polar coordinate system may be used to program the locations for holes in a bolt circle. By locating the bolt circle center, using the Local Coordinate System function G52, a rotation angle and circle radius can be programmed to locate the holes (unless the center is X0Y0). The programmed values in the X axis represent the circle radius and, in the Y, represent the angular. The angular values may be programmed as either positive, or negative. Positive rotation is counterclockwise and negative rotation is clockwise.

Plane Selection (G17, G18, G19)

G17 = XY plane

G18 = XZ plane

G19 = YZ plane

By declaring the selection of plane G17, G18, and G19, the machine automatically selects the given plane. The default plane on a vertical machining center is the G17 XY plane and, therefore, it is not necessary to input G17 (but it should be part of the safety block).

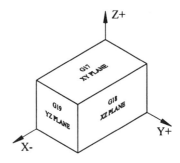

Figure 25

Example of Program No. 7

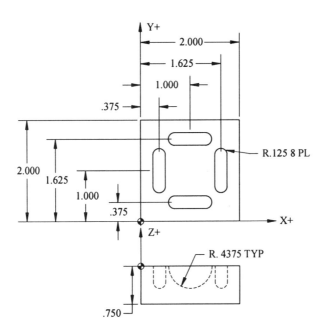

Figure 26

225

Part 4 Machining Center Preparatory Functions

| Machine: Machining Center | Program Number: O0007 |

Workpiece Zero X = <u>Lower Left Corner</u> Y = <u>Lower Left Corner</u> Z = <u>Top Surface</u>

Setup Description:

Material used is aluminum 6061-T6 $S = \frac{12 \times V}{\pi \times d} = 4585$, Feed per flute .001;

$F = 2 \times .001 \times 4585 = 9.17$

Tool #	Description	Offset	Comments
T01	.250 diameter HSS 2 flute ball end mill		SFM = 300 Feed =.001 inch per flute

Program

O0007

N10G90G20G17G80G40G49

N15G92X10.0Y7.0Z0

N20G90G00X1.3125Y.375S4585M03

N25G43Z1.H01M08

N30Z.2

N35G01Z-.125F18.0

N40G18G03X.6875Z-.125I-.3125K0F9.0

N45G01Z.2F18.0

N50G00Y1.625

N55G01Z-.125

N60G02X1.3125Z-.125 I.3125K0F9.0

N65G01Z.2F12.0

N70G00X1.625Y1.3125

N75G01Z-.125

N80G19G02Y.6875Z-.125J-.3125K0F9.0

N85G01Z.2F18.0

N90G00X.375

N95G01Z-.125

N100G03Y1.3125Z-.125J.3125K0F9.0

N105G01Z.2

N110G91G28Z0M09

N115G28X0Y0M05

N120M30

Part 4   Machining Center Preparatory Functions

Input In Inches (G20), Input In Millimeters (G21)

Function G20 or G21 is entered at the beginning of the program to establish the measurement system. They apply to the whole program. Functions G20 and G21 cannot be interchanged during programming. When using functions G20 or G21 the values are the same units of measure for:

F = feed rate

Position of X, Y, and Z

Offset values.

The default measurement system for most American machines is inch, therefore, it is not necessary to use the above functions unless the opposite system is required.

Stored Stroke Limit (G22, G23)

Function G22 identifies the stored stroke limit for the tool area outside the working envelope. Function G23 identifies the stored stroke limit for the tool area inside the working envelope.

The manufacturer enters these stroke limits of the axis travel into the controller memory, in order to define the work envelope for the machining center. In case the tool is positioned beyond the limited area (outside the limits of the stroke), the machine will stop and an alarm is displayed on the screen.

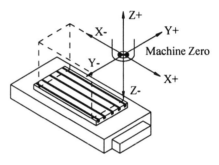

Figure 27

Reference Point Return Check (G27)

This function checks for the accurate return to the reference point and confirms the programmed position has been reached. Entering function G27 causes rapid traverse of the tool to the reference point. If the tool is positioned accurately at the reference point, the axis LED's will light up. If the LED's do not go on, the tool is displaced from the reference point and an alarm will result. This might occur, if prior to the G27 command, cancellation of the tool radius compensation has been omitted (i.e., function G40 is omitted following the use functions G41 and G42). This will cause a position displacement of the tool equal to the value of the compensation offset and the execution of the following program block with the same displacement.

Return To The Reference Point (G28)

This function is the automatic reference point return through an intermediate point programmed in the X, Y and/or Z axes. The machine will position at rapid traverse (G00) to the programmed intermediate point coordinate values and then to the reference point of machine zero. This function is commonly used before an automatic tool change (ATC). After using function G28, the axis LED's on the control panel should light up to indicate

Part 4 Machining Center Preparatory Functions

successful return to zero for the axis. When you use function G28, you must specify the point through which the tool passes on its way to zero. If the command G28 X0 Y0 Z0 is entered in the incremental mode (G91), the machine will position at rapid traverse to the reference position in all three axes simultaneously. Caution must be exercised so as not to interfere with the work holding device. The following example demonstrates the best method for using function G28 by first positioning the Z axis to a level for clearance of the work holding.

Example:

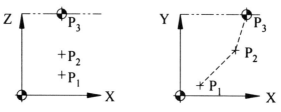

Figure 28

Function G90 is active.

G28 Z3.00

G28 X6.00 Y7.00

where

X, Y, and Z = the coordinates of the point through which the tool passes on its way to zero return

P_1 = present position

P_2 = point through which the tool passes

P_3 = machine home position

Return From The Reference Point (G29)

To use function G29 it must follow function G28 or G30. It can be applied after an automatic tool change so that the tool will return to the work position at rapid traverse, as it similarly does in G00. The tool will travel through the intermediate point programmed.

Example:

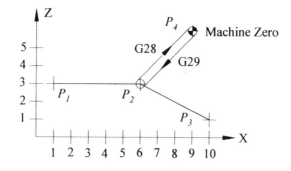

Figure 29

Part 4 Cutter Diameter Compensation

G91 = incremental positioning command

    G28X5.0Y0 P_1 - P_2 - P_4

    M06

    G29X4.0Y-1.5 P_4 - P_2 - P_3

where

P_1 = initial work point of tool 1

P_2 = end work point of tool 1

P_2 = initial work point of tool 2

P_3 = end work point of tool 2

P_4 = machine home

During the execution of the block containing function G29, the tool automatically follows the return path from point P_1 to P_2 and the programming is limited only to entering the difference between points P_2 and P_3.

Return To Second, Third, And Fourth Reference Points (G30)

Usually machine zero coordinates are coincidental to the reference point assigned to function G28. However, if the reference point assigned to function G28 does not correspond to machine zero, function G30 must be used in order to place the tool in this position. (This may be the case if the tool change position is not the same point as is specified in function G28.) Before applying function G30, apply function G28. The movement determined by function G30 is rapid traverse. Functions G28, G29, and G30 can be used to program in both absolute and incremental systems. Cutter compensations (length and diameter) must be cancelled prior to this command.

Cutter Compensation (G40, G41, G42)

    G40 = Cutter radius compensation cancel

    G41 = Cutter radius compensation left

    G42 = Cutter radius compensation right

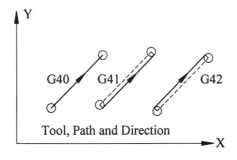

Figure 30

Part 4   Cutter Diameter Compensation

The use of cutter compensation allows the programmer to use the part geometry exactly as from the print for programmed coordinates. Without using compensation, the programmer must always know the cutter size and offset the programmed coordinates for the geometry by the amount of the radius. In this scenario, if a different size cutter is used the part will not be machined correctly.

An added advantage for using cutter compensation is the ability to use any size cutter as long as the offset amount is input accurately into the offset register.

It is also very effectively used for fine-tuning of dimensional results by minor adjustments to the amount in the offset register.

Cutter Compensation Cancellation (G40)

Function G40 is used to cancel cutter radius compensation initiated by G41 or G42. It should be programmed after the cut using the compensation is completed by moving away from the finished part in a linear (G01) or rapid traverse (G00) move by at least the radius of the tool. Care should be take here because if the cancellation is on a line without movement, the cutter will move in the opposite direction and may damage the part.

Cutter Radius Compensation Left and Right (G41 and G42)

Functions G41 and G42 offset the programmed tool position to the left (G41) or right (G42), by the value of the tool radius entered into offset registers and called in the program by the letter address D. For each tool, enter the corresponding offset amount. In the program, the letter D and the number of the offset (two digits) is input to initiate the compensation call.

The direction the tool is offset, to the left or the right, is dependant upon which direction the tool is traveling. To accomplish climb cutting with right hand tools, always use G41 and for conventional cutting use G42. Consider which direction of offset is needed by facing the direction the tool is going to travel and observe which side of the part the cutter is going to be on, to the left for G41 or to the part or right for G42.

Procedure For Initiating Cutter Compensation

Position the tool in the X and Y axes to a point away from the required finished geometry. Then program a linear move that is larger than the radius of the cutter to feed into the part. i.e. G01 G41 or G42 offset direction X or Y absolute coordinate and D offset number. To use the cutter compensation properly, there needs to be one full line of movement to position the tool on the proper vector to cut the part. Once this is accomplished, program the part geometry per print.

The tool will not be positioned to the actual programmed point on the geometry, rather, it will be positioned to a point plus or minus the offset value of the cutter called by D, and the edge of the cutter will be aligned with finished part geometry.

Rules For Cutter Compensation Use

When cutter compensation is called in a program, the control looks ahead by two lines in order to set up a vector to position for the offset amount for each move.

Part 4 Cutter Diameter Compensation

Movement must be maintained once G41 or G42 commands have been called up. By following this rule, over-cutting of the part can be avoided. If two lines of non-movement commands are placed in a row after cutter compensation is called up, the control will ignore functions G41 or G42 and the part will be cut incorrectly.

Do Not start compensation G41 or G42 when G02 or G03 is in effect.

When a change from left to right compensation, or vise versa is needed, it is recommended that cancellation of the first compensation be executed, then the second compensation may be called and; thus, the transition of the tool position vector will not conflict.

When machining an inside radius, the radius must be larger than the offset of the tool, or the control will stop the program and an alarm will be displayed.

The move used to call up cutter compensation must be larger than the radius of the tool used.

Example of Program No. 8

Offset for tool number one is D01.

Offset for tool number two is D02.

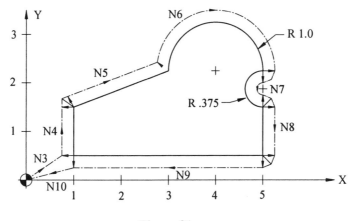

Figure 31

Please note in the following program example there are no Z axis moves. The intent here is to demonstrate the usage of cutter radius compensation only. The program line sequence numbers in illustration above correspond with the following program. The arrows along the tool centerline path indicate the travel direction. At each intersection point on the geometry the arrows indicate the tool offset direction.

Program

N1G90G17G20G80G40G49

N2G92X7.0Y-5.0Z0

N3G01G41X1.00Y.50D01F10.0

Part 4 Cutter Diameter Compensation

N4G01Y1.5

N5G01X3.00Y2.25

N6G02X5.00Y2.25I1.00

N7G03X5.00Y1.50J-.375

N8G01X5.00Y.50

N9X1.00

N10G00G40X0Y0

N11M05

N12M30

Notes: At the end of the tools work, function G40 must be applied to cancel any previously entered compensation value of offset D.

Example:

G41D01G02X3.Y-3.J-3.

O = programmed center of an arc

O' = displaced center of an arc

Using functions G41 and G42, in the same block along with functions G02 and G03, causes displacement of the center of an arc by the amount in the offset register. This action will result in a controller alarm.

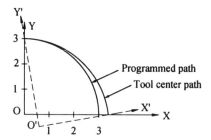

Figure 32

Radial Offset Vector

The radial offset vector is defined as the distance between the center of the tool and the programmed contour. The offset vector is always perpendicular to the programmed line or arc. The offset vector is two-dimensional (for two axes) and depends on the choice of plane (G17, G18, or G19). The following examples demonstrate the tool path taken when the execution of functions G41 and G42 are initiated and then cancelled by function G40. With angles of the machined part, $\alpha \le 180°$ two types (A and B) of tool approaches, to and from the machined part can be found when using functions G41 and G42. Types A or B are permanently stored in the machine's control parameters.

Type A: $90° \le \alpha \le 180°$

Figure 33 Line–line

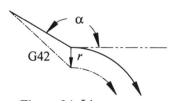

Figure 34 Line–arc

Part 4 Cutter Diameter Compensation

Figure 35 Line–line

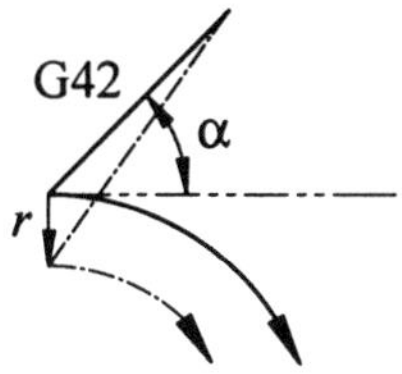

Figure 36 Line–arc

Type B: $90° \leq \alpha \leq 180°$

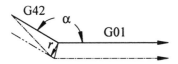
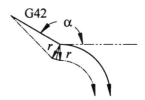

Figure 37 Line–line

Figure 38 Line–arc

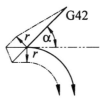

Figure 39 Line–line

Figure 40 Line–arc

In type B, when the angle is less than 1°:

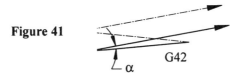

Figure 41

The following are examples of tool withdrawal from the workpiece when using tool radius compensation cancel function G40.

Type A: $\alpha < 90°$

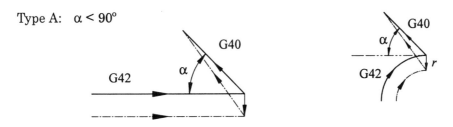

Figure 42 Line–line

Figure 43 Line–arc

Part 4 Cutter Diameter Compensation

$90° \leq \alpha \leq 180°$

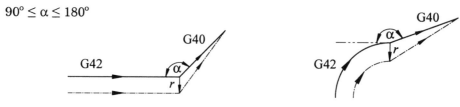

Figure 44 Line–line

Figure 45 Arc-line

Type B: $\alpha < 90°$

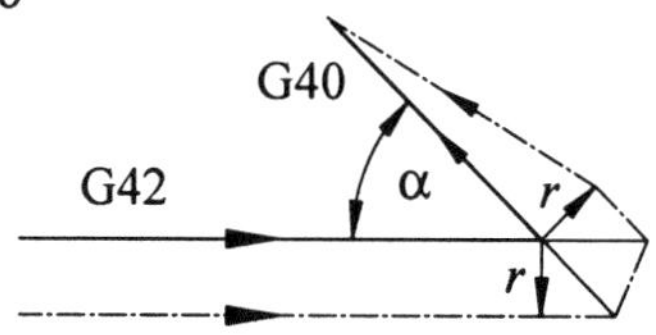

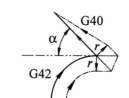

Figure 46 Line–line

Figure 47 Arc-line

$\alpha < 1°$ **Figure 48**

$90° \leq \alpha \leq 180°$

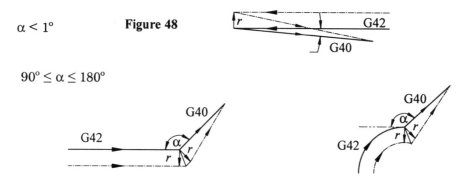

Figure 49 Line–line

Figure 50 Arc-line

The following is an example of a tool path configuration with reference to the program line, when executing functions G41 and G42.

$\alpha < 90°$ $\alpha < 1°$

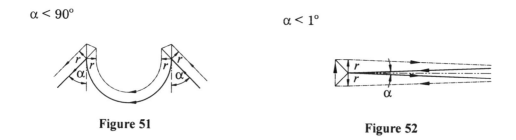

Figure 51 **Figure 52**

Part 4 Cutter Diameter Compensation

$90° \leq \alpha \leq 180°$ $\alpha > 180°$

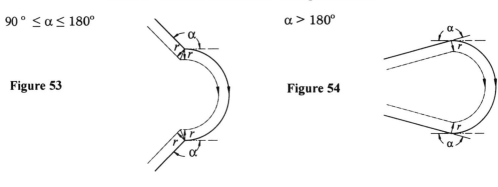

Figure 53   **Figure 54**

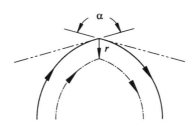

Figure 55

Special Cases:

The radius of the machined workpiece is smaller than the tool radius.

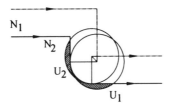

Figure 56

The control has a look-ahead feature that always reads one or more lines of the program in advance. For the above, at the end of the execution of block N_1, an alarm will be displayed on the screen and the current job will be interrupted. If the execution of the program proceeds line by line (the SINGLE BLOCK button is ON) after the completion of operation N_1, the control will not read the information contained in N_2 and thus cannot offset the tool path to compensate for the condition and will undercut the edge in U_1. A similar situation will be encountered with undercut U_2.

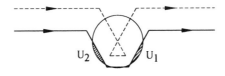

Figure 57

Part 4 Cutter Diameter Compensation

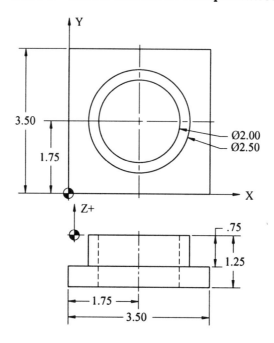

Figure 58

| Machine: Machining Center | Program Number: O0009 |

Workpiece Zero X = <u>Lower Left Corner</u> Y = <u>Lower Left Corner</u> Z = <u>Top Surface</u>

Setup Description:

Material used is 3.5 × 3.5 × 1.25, 303 Stainless Steel.

Tool #	Description	Offset	Comments
T01	#5 HSS CTR drill		SFM = 100 Feed =.002 inch per flute
T02	1.25 diameter drill		SFM = 100 Feed =.003 inch per flute
T03	1.25 diameter, six flute, HSS roughing cutter	D60 = .650	SFM = 100 Feed =.003 inch per flute
T04	1.25 diameter, six flute, HSS end mill	D61 = .625	SFM = 100 Feed =.002 inch per flute

Example of Program No. 9

Program

O0009

N10G90G20G80G40G49

N15G92X10.0Y7.0Z0

N20T01M06

N25G00X1.75Y1.75S915M03T02

Part 4 Cutter Diameter Compensation

N30G43Z.1H01M08

N35G81G98Z-.382R.1F4.0

N40G00G80Z1.0M09

N45G91G28Z0

N50M01

N55T02M06

N60G90G80G40G49

N65G00X1.75Y1.75S320M03T03

N70G43Z.1H02M08

N75G83G98Z-1.625R.1Q.625F2.0

N80G00G80Z1.0M09

N85G91G28Z0

N90M01

N95T03M06

N100G90G80G40G49

N105G00X1.75Y1.75S320M03T04

N110G43 Z.1H03M08

N115 1Z-1.3F20.0

N120G01G42Y2.750D60F5.76

N125G02 J-1.0

N130G01 G40Y1.750F10.0

N135G00Z.1

N140Y-1.0

N145G01Z-.73F20.0

N150G42Y.5D60

N155G03 J1.25F6.0

N160G01G40Y-1.0 F20.0

N165G00 Z.1M09

N170G91G28Z0

N175M01

N180T04M06

N185G90 G80G40G49

Part 4 Cutter Diameter Compensation

N190G00X1.75Y1.75S320M03T01

N195G43Z.1H04M08

N200G41Y2.75F10.D61

N205G01Z-1.3F50.0

N210G03J-1.0F3.84

N215G01G40Y1.75F20.0

N220G00Z.1

N225Y-1.0

N230Z-.75

N235G01G41Y0.50F10.D61

N240G02J1.25F4.0

N245G01G40Y-1.0F20.0

N250G00Z.1M09

N255G91G28Z0M05

N260G28X0Y0

N265M30

Example of Program No.10

The following is an example illustrating the machining of a slot through a material 0.5 inch thick.

Machine: Machining Center		Program Number: O0010	
Workpiece Zero X = <u>Lower Left Corner</u> Y = <u>Lower Left Corner</u> Z = <u>Top Surface</u>			
Setup Description:			
Tool #	Description	Offset	Comments
T01	#5 HSS CTR drill		SFM = 300 Feed =.003inch per flute
T02	$\frac{17}{32}$ diameter HSS drill		SFM = 300 Feed =.003 inch per flute
T03	$\frac{1}{2}$ diameter, two flute, HSS roughing cutter	D50 = .260	SFM = 850 Feed =.003 inch per flute
T04	$\frac{1}{2}$ diameter, two flute, HSS end mill	D51 = .250	SFM = 850 Feed =.002 inch per flute

Part 4 Cutter Diameter Compensation

Program

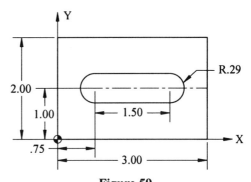

Figure 59

O0010

N10G90G80G20G40G49

N15T01M06

N20G92X10.0Y7.0Z0

N25G00X.750Y1.S2200M03T02

N30Z.1H01M08

N35G81G98Z-.3R.1F8.8

N40G80Z1.0M09

N45G91G28Z0

N50M01

N55T02M06

N60G90G80G20G40G49

N65G00X.75Y1.0S2150M03T03

N70G43Z1.H02M08

N75G81G98Z-.70R.1F8.0

N80G80Z1.0M09

N85G91G28Z0

N90M01

N95T03M06

N100G90 G80G40G49

N105G00X.75Y1.0S2200M03

N110G43 Z1.0H03M08

N115Z.1

N120G01Z-.52F50.0

N125G41Y.71F4.5D50

N130X2.25

N135G03Y1.29J.29

N140G01X.75

N145G03Y.71J-.29

N150G01G40Y1.0

Part 4 Cutter Diameter Compensation

N155G00Z1.0M09

N160G91G28Z0

N165M01

N170T04M06

N175G90G80G40G49

N180G00X.75Y1.0S2200M03T05

N185G43Z1.0H04M08

N190Z.1

N195G01Z-.52F50.0

N200G41Y1.29F8.8D51

N205G03Y.71J-.29

N210G01X2.25

N215G03Y1.29J.29

N220G01X.75

N225G40Y1.0

N230G00Z1.0M09

N235G91G28Z0M05

N240G28X0Y0

N245M30

Tool Length Compensation (G43, G44, G49)

G43 = positive tool length offset

Note: the entry in the offset register is negative because the offset represents the distance from the tool tip to the workpiece Z zero.

G44= negative tool length offset (not often used today)

G49 = cancellation of tool length offset

Functions G43 and G44 are used to read the tool length offset amount from the offset registers. Numbers in the offset registers correspond to the tool length in the zero position from the surface of the machined part. The tool length offset is called in the program by the letter address H. The tool length offset value may be added (G43) to the programmed value of Z or subtracted (G44) from the programmed value of Z.

The tool offset number used is identified by the letter H and two digits.

Example: H01 = offset number one

Part 4   Work Coordinate Systems

When entering the new offset value of H to the offset register, the previous offset value is automatically canceled and the machine reads the new value without considering the previous value. To cancel functions G43 and G44, use function G49.

Offset Amount Input By The Program (G10)

If the number of offsets exceeds the limit of the machine's computer memory containing the offset for the tool, tool length compensation, or tool radius compensation, then an additional offset may be entered directly into the program.

Example:

G10P...R...

where

P = offset number

R = value of offset

If all the work coordination systems (functions G54 through G59) are reserved, then entering the new coordination values (X, Y, and Z) in the program, along with the function G10, six additional systems are gained.

Example:

G10L2 P6 X...Y...Z...

In the case of the sixth coordinate system (G59), the previous coordinate values (X, Y, and Z) are replaced by new ones.

G10L2P1:G54

G10L2P2:G55

G10L2P3:G56

G10L2P4:G57

G10L2P5:G58

G10L2P6:G59

Note: The function G10 is used to change the work coordination values and for other offsets. Therefore, if you plan to use the new coordination system, call directly for the newly entered system in the consecutive block (in the example, G59). Before G10, you should use function G90.

This G10 format varies with control models, consult the specific manufacturer control manuals.

Work Coordinate Systems (G54, G55, G56, G57, G58, G59)

In comparison to function G92 (programming absolute zero at the origin of the coordinate system), the application of functions G54 through G59 simplifies the programming where it is advisable to use additional workpiece zeros (maximum six as a standard control feature).

Part 4 Work Coordinate Systems

Note: When using functions G54 through G59, avoid using function G92. Using function G92 with function G54 may cause unpredictable problems.

Example: First (G54) and second (G55) coordinating systems

While using function G92, the coordinates of workpiece zero are entered directly into the program. When using functions G54 through G59, the coordinates of the program zeros are entered into controller memory on the work offset page of the coordinate systems. Under the position 01, enter the coordinates of workpiece zero in the function G54, and so on.

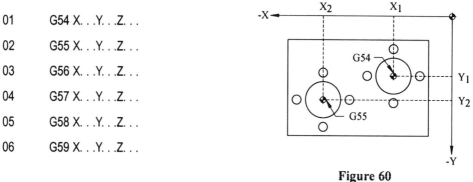

01	G54 X...Y...Z...
02	G55 X...Y...Z...
03	G56 X...Y...Z...
04	G57 X...Y...Z...
05	G58 X...Y...Z...
06	G59 X...Y...Z...

Figure 60

Workpiece zero coordinates G54 through G59 are measured in a similar manner as they are for function G92. Function G92 is measured from X Y zero to the tool tip while G54 through G59 is measured from machine zero to part zero.

The following two examples illustrate the application of the functions G92 and then G54 through G59:

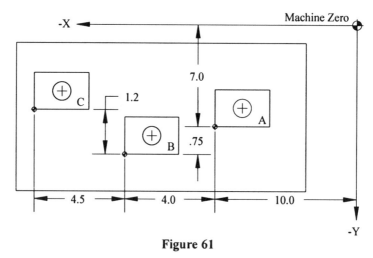

Figure 61

Example of Program No.11

In the following program, the tool used is a .5 diameter HSS drill and cutting speed of 80 ft/min was used to calculate the RPM and feed rate.

Part 4 Work Coordinate Systems

The following is a program utilizing function G92:

O0011

N10G90G80G20G40G49

N15G92X10.0Y7.0Z0 (PART A)

N20G00X.75Y.5S611M03

N25G43Z1.0M08H01

N30G00Z.1

N35G01Z-.65F3.6

N40G00Z.1

N45X0Y0

N50G92X14.0 Y7.75 (PART B)

N55G00X.75Y.5

N60G01Z-.65F3.6

N65G00Z.1

N70X0Y0

N75G92X18.5Y6.55 (PART C)

N80G00X.75Y.75

N85G01Z-.65F3.6

N907G00 Z.1

N95G28Z1.0M09

N100G91X0Y0M05

N105M30

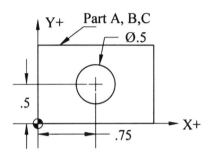

Figure 62

The following is a program utilizing the functions G54 through G59:

Part No. 1: (A) G54   X–10.0 Y–7.0   Z0
Part No. 2: (B) G55   X–14.0 Y–7.75   Z0
Part No. 3: (C) G56   X–18.5 Y–6.55   Z0

O0011

N10G9080G20G80 G40G49

N15G00G54X.75Y.5S611M03

N20G43Z1.0M08H01

N25G00Z.1

Part 4 Work Coordinate Systems

N30G01Z-.65F3.6

N35G00Z.1

N40G55G00X.75 Y.5

N45G01Z-.65

N50G00Z.1

N55G56G00X.75Y.5

N60G01Z-.65

N65G00Z.1

N70G28Z1.0M09

N75G91G28X0Y0M05

N80M30

Reviewing the above examples, you will note the difference between the function G92 and the coordinate systems G54 through G59.

Single-Direction Positioning (G60)

Usually, function G60 is applied when accuracy is the prime factor in determining distances between points. The tool will always approach the programmed point from one direction only.

The value of the additional path is set by parameter in the control. Function G60 eliminates the undesirable influence of gear and feed-screw play (backlash) on the accurate positioning of the programmed point.

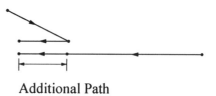

Additional Path

Figure 63

Canned Cycle Functions

The function of a given cycle is defined as a set of operations assigned to one block and performed automatically without any possibility of interruption. Usually, it is a set of six operations, as follows:

1. Positioning of the X and Y axes at rapid traverse.

2. A rapid traverse move to an initial clearance (R) level plane (G98).

3. The machining cycle is executed (drill, bore, etc.).

4. A dwell or other operation is executed at the bottom of the hole.

5. A rapid traverse return to the R level plane along the Z axis (G99).

6. A rapid traverse return to the initial level plane along the Z axis.

Part 4 Canned Cycle Functions

Block notation is as follows:

N...G...G...X...Y...Z...R...Q...P...F...K...L...

Where

N = the block number

G = the type of cycle function

G = initial or R level return G98/G99

X, Y = the hole position (usually positioning is carried out by rapid traverse)

Z = the depth of the hole

R = the distance between plane R and the surface of the material

Level R refers to the horizontal plane, positioned closely above the material on which the tool tip moves. The programmed value of R is valid until the new value is entered. It does not have to be included in every block. The tool will return to this level at the end of each hole drilled.

Q = the depth of cut (drilling) for individual pecks (not used for every drilling cycle)

P = the dwell time for the drill while rotating at the bottom of the hole (not used for every drilling cycle)

The dwell in seconds is for the purpose of accurate removal of excess material.

F = feed rate in inches per minute (IPM).

K = the number of repeats

L = the number of holes incrementally spaced

When K is used with G91, L represents the number of holes incrementally spaced by the amount entered as the X or Y position coordinate (if L does not appear in the block, that means machining of only one hole, L = 1). On some modern controls, the letter address K is used in the same manner.

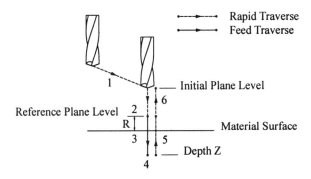

Figure 64

Part 4 Canned Cycle Functions

1. Positioning with rapid traverse
2. Rapid traverse to level R along Z axis
3. Feed traverse to Z depth
4. Operations performed on the bottom of the hole
5. Return to level R (G99)
6. Return to plane level (G98)

High Speed Peck Drilling Cycle (G73)

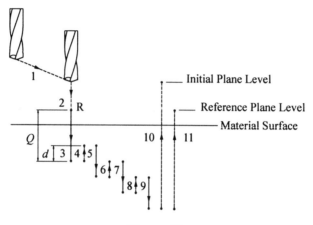

Figure 65

G73X...Y...Z...R...Q...F...

Figure explanations:
1. G00 — rapid traverse to a position in X or Y axes.
2. G00 — rapid traverse to plane R.
9, 7, 5, 3. G01 — feed traverse along Z to depth Q.
8, 6, 4. G00 — rapid traverse upwards by a value specified.
Established by parameter and shown here as amount "d".
10. G00 — rapid traverse to a plane level assigned to function G98.
11. G00 — rapid traverse to plane R assigned to function G99.

Q = The depth of cut for each drilling peck.

d = The value of the upward traverse that is used to accomplish momentary interruption, removal of chips, and the delivery of coolant to the bottom of the hole. The value of d is entered into the parameters of the machine's control.

Part 4 Canned Cycle Functions

Left Handed Tapping Cycle (G74)

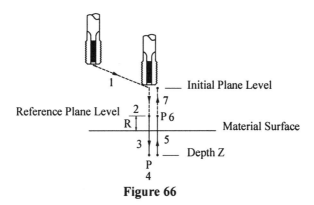

Figure 66

G74 X...Y...Z...R...P...F...

The spindle is rotating in the counterclockwise direction by using the M04 command with a value of S.

1. G00 — rapid traverse to a position in X or Y axes.
2. G00 — rapid traverse to plane R.
3. G01 — feed traverse along Z to depth.
   P. — P = dwell time in seconds at the bottom of the hole
4. — spindle direction is reversed to clockwise.
5. G01 — feed traverse along the Z to the R plane.
   P. — P = dwell time in seconds at the R plane
6. — spindle direction is reversed to counterclockwise.
7. G00 — rapid traverse to initial plane if G98 is used.

Fine Boring Cycle (G76)

G76 X...Y...Z...R...Q...P...F...

1. G00 — rapid traverse to point R.
2. G01 — feed traverse to Z depth.
   P. — dwell time in seconds
3. M19 — miscellaneous function M19.

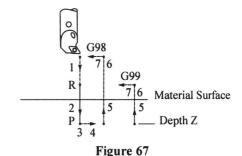

Figure 67

Miscellaneous function M19 initiates orientation of the spindle position and the interruption of revolutions. At function M19, the cutting edge always rotates and stops in the same position. The cutting edge stops so that it is perpendicular to the X or Y axis. This activity takes place without inserting the miscellaneous function M19 into the program. It is a part of the G76 canned cycle that is built-in.

Part 4 Canned Cycle Functions

Example:

4. G00 — rapid displacement along the X or Y axis

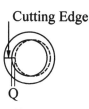

Figure 68

The displacement may be in the Y axis, if the cutting edge is perpendicular to the X axis, with the value of Q. The value of Q is entered into the parameters of the control. The value of Q must be known in order to avoid a collision between the tool and the back wall of the hole.

5. G00 — rapid traverse from the hole to the reference point R for function G99, or rapid traverse to the initial plane level for function G98

6. G00 — rapid displacement of the tool along the X (or Y) axis, with the value Q

7. — M03 activates clockwise spindle rotation; this is in preparation for the boring of the next hole

This boring cycle is usually applied for finishing in which it is intended to obtain a smooth surface, free of scratches. Because of the retract amount specified in Q, there will be no tool mark caused on the finished surface at exit from the hole.

Canned Drilling Cycle Cancellation (G80)

This command is used to cancel all canned cycles. It should be entered at the end of each canned cycle machining sequence. This code is also entered in the safety block.

Notes for cycles G73 through G89:

Revolutions of the spindle (right or left), during the cycle, must be entered in the block proceeding the canned cycle block.

Never press the ORIGIN button (zero set) during the canned cycle execution because unpredictable actions may result.

In order to execute the cycle (the drilling of one hole with the SINGLE BLOCK button being ON, after the execution of each block when the machine stops), you must press the CYCLE START button three times.

If feed hold is pressed during a canned cycle operation, the cycle will be completed before stopping.

Drilling Cycle, Spot Drilling (G81)

G81X...Y...Z...R...F...

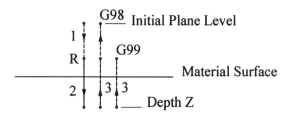

Figure 69

Part 4 Canned Cycle Functions

1. G00 — rapid traverse to R.
2. G01 — feed traverse to Z depth.
3. G00 — rapid traverse to R (G99), or
4. G00 — rapid traverse to initial level plane (G98).

Example of Program No.12

Drilling a hole with a diameter of .25 and a chamfer with a diameter of .280.

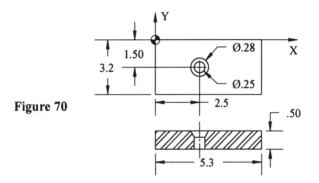

Figure 70

Machine: Machining Center	Program Number: O0012

Workpiece Zero X = <u>Upper Left Corner</u> Y = <u>Upper Left Corner</u> Z = <u>Top Surface</u>

Setup Description:

The material is 1018 low carbon steel.

Mount the workpiece in a vice, with a part stop on the left side. Parallels are inserted below the surface of the plate on the side of the vice jaws (dotted line on the drawing).

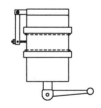

Figure 71

Tool #	Description	Offset	Comments
T01	#4 HSS CTR drill		SFM = 110 Feed =.002 inch per flute
T02	.250 diameter HSS drill		SFM = 110 Feed =.0035 inch per flute

Program

O0012

N10G90G80G20G40G49

N15T01M06

N20G54G00X2.5Y-1.5S1408M03T02

N25G43Z1.H01M08

Part 4 Canned Cycle Functions

N30G81G98Z-.28R.1F6.0

N35G80Z1.M09

N40G91G28Z0

N45M01

N50M06

N55G90G80G40G49

N60G00X2.5Y-1.5S160M03

N65G43Z1.H02M08

N70G81G98Z-.58R.1F12.32

N75G80Z1.M09

N80G91G28Z0

N85G28X0Y0M05

N90M30

Counter Boring Cycle (G82)

In function G82, the feed is interrupted in the drilling cycle (while the spindle is ON) at the bottom of the hole for time specified by P.

    G82X...Y...Z...R...P...F...

1. G00 — rapid traverse to reference level R.
2. G01 — feed traverse to Z depth.
3. — interruption of the feed for time duration P, in order to remove the material at the bottom of the hole.
4. G00 — rapid traverse to initial plane level for G98.
5. G00 — rapid traverse to reference plane level R for G99.

Example of Program No.13

Drilling of a counter bored (step hole) bolt hole.

Figure 72

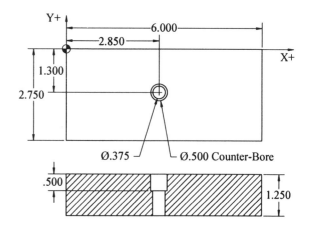

Part 4 Canned Cycle Functions

| Machine: Machining Center | Program Number: O0013 |

Workpiece Zero X = <u>Upper Left Corner</u> Y = <u>Upper Left Corner</u> Z = <u>Top Surface</u>

Setup Description:

The material is 1018 low carbon steel.

Fasten the workpiece in the same way as in the previous example.

Tool #	Description	Offset	Comments
T01	#4 HSS CTR drill		SFM = 110 Feed =.002 inch per flute
T02	.375 diameter HSS drill		SFM = 110 Feed =.0035 inch per flute
T03	.500 diameter HSS, 2 flute end mill		SFM = 110 Feed =.002 inch per flute

Program

O0013

N10G90G80G20G40G49

N15T1M06

N25G54G00X2.85Y-1.3S1408M03T02

N30G43Z1.H01M08

N35G81G98Z-.4R.1F5.63

N40G80Z1.M09

N45G91G28Z0

N50M01

N55M06

N60G90G80G40G49

N65G00X2.85Y-1.3S1173M03T03

N70G43Z1.H02M08

N75G81G98Z-1.4F8.21R.1

N80G80Z1.M09

N85G91G28Z0

N90M01

N95M06

N100G90G80G40G49

N105G00X2.85Y-1.3S880M03T01

N110G43Z1.H03M08

Part 4 Canned Cycle Functions

N115G82G98Z-1.4R.1P200F7.04

N120G80Z1.M09

N125G91G28Z0

N130G28X0Y0M05

N135M30

Note: On some controls, if no value is assigned for P to function G82, its value will automatically be selected by the control. If you do enter a value for dwell P (for example, P1000 = 1 second), then the constant value included in the parameters of the machine will be ignored.

Deep Hole Peck Drilling Cycle (G83)

G83X...Y...Z...R...Q...F...

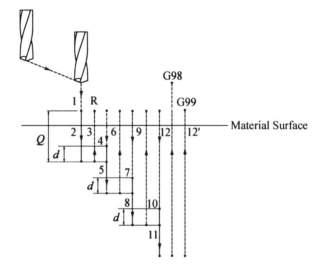

Figure 73

1. G00 — rapid traverse to R level.
2. G01 — feed traverse with length Q.
3, 6, 9. G00 — rapid return traverse to R.
4, 7, 10. G00 — rapid traverse to the depth previously drilled, less the value of d.
5, 8, 11. G01— feed traverse increased by the value of d.
12. G00 — rapid traverse to initial plane level for G98.
12'. G00 — rapid traverse to reference plane level R for G99.

This drilling cycle is used for exceptionally deep holes, because during each work traverse, the drill returns, at rapid traverse, to the R level point, which allows the removal of the chips and the delivery of the coolant to the bottom of the hole. Entering a given depth

Part 4 Canned Cycle Functions

of Z into the control enables it to calculate the number of feed traverses necessary for *Q*. *Q* can be any incremental amount desired smaller than the total Z axis travel. The value of *Q* does not need to have a common factor with the dimension Z. The value of *d* is set by parameter.

Example of Program No.14

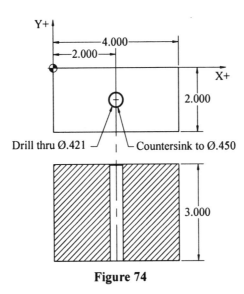

Figure 74

Machine: Machining Center	Program Number: O0014

Workpiece Zero X = <u>Upper Left Corner</u> Y = <u>Upper Left Corner</u> Z = <u>Top Surface</u>

Setup Description:

The material is 1018 low carbon steel.

Fasten the workpiece in the same way as in the previous example.

Tool #	Description	Offset	Comments
T01	#5 HSS CTR drill		SFM = 110 Feed =.002 inch per flute
T02	.421 diameter HSS drill		SFM = 110 Feed =.0035 inch per flute

Program

O0014

N10G90G80G20G40G49

N15T1M06

N20G54G00X2.Y-.1S1006M03T02

N25G43Z1.H01M08

Part 4 Canned Cycle Functions

N30G81G99Z-.38F4.03R.1

N35G80Z1.M09

N40G91G28Z0

N45M01

N50M06

N55G90G80G40G49

N60G54G00X2.Y-1.S1045M03

N65G43Z1.H02M08

N70G83G99Z-3.15R.05Q.45F7.31

N75G80Z1.M09

N80G91G28Z0

N85G28X0Y0M05

N90M30

Note: By using function G83 in block N70, the deep hole, peck drilling cycle is initiated. Starting at level R, the drill will feed by the amount of Q.45 and then, at rapid traverse, returns to the starting point R. The next move advances by the depth Q, decreased by the value of d, which is entered in the parameters of the control. This cycle repeats itself until the drill reaches a depth of Z-3.15. Also, after each feed traverse with the value Q, the tool returns to level R.

Tapping Cycle (G84)

G84 X...Y...Z...R...P...F...

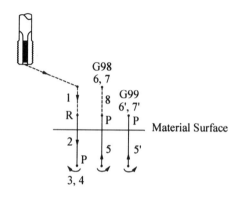

Figure 75

254

Part 4 Canned Cycle Functions

1. G00 — rapid traverse to R level.
2. G01 — feed traverse to Z.
3. M05 — revolutions stop.
P. — dwell time at the bottom of the hole.
4. M04 — counterclockwise revolutions are ON.
5. G01 — feed traverse to R.
6. M05 — spindle stop.
7. M03 — clockwise revolution is ON.
8. G00 — rapid traverse to a level plane for the function G98.

Example of Program No.15

Threading cycle

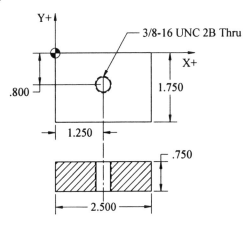

Figure 76

Machine: Machining Center	Program Number: O0015

Workpiece Zero X = Upper Left Corner Y = Upper Left Corner Z = Top Surface

Setup Description:

The material is 1018 low carbon steel.

Fasten the workpiece in the same way as in the previous example.

Tool #	Description	Offset	Comments
T01	#4 HSS CTR drill		SFM = 110 Feed =.002 inch per flute
T02	$\frac{5}{16}$ diameter HSS drill		SFM = 110 Feed =.0035 inch per flute
T03	$\frac{3}{8}$ – 16 Tap		SFM = 15

Part 4 Canned Cycle Functions

Program

O0015

N10G90G80G20G40G49

N15T1M6

N20G00G54X1.25Y-.8S1045M03T02

N25G43Z1.H01M08

N30G81G99Z-.38R.1F4.18

N35G80Z1.M01

N40G91G28Z0

N45M01

N50M06

N55G80G40G49

N60G54G00X1.25Y-.8S1408M03T03

N65G43Z1.H02M08

N70G81G99Z-.85R.1 F9.85

N75G80Z1.M09

N80G91G28Z0

N85M01

N90M06

N95G80G40G49

N100G54G00Z1.25Y-.8S152M03T01

N105G43Z1.H03M08

N110G84G99Z-1.R.1F9.5

N115G80Z1.M09

N120G91G28Z0M05

N125G28X0Y0

N130M30

Note: The feed in block N110 (F9.5) was calculated as follows:

$$F = 1/16 \times 152 = 9.5$$

1/16= the lead of the thread in inches

152 = the spindle speed, expressed in revolutions per minute

Part 4 Canned Cycle Functions

Boring Cycles

G85 Reaming Cycle

G85 X ... Y ... Z ... R ... F ... K ...

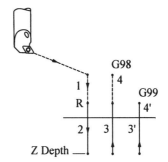

Figure 77

1. G00 — rapid traverse to R.
2. G01 — feed traverse to Z depth.
3. G01 — feed traverse to the R level plane and then rapid to the level plane assigned to function G98.
3'. G01 — feed traverse to R for function G99.
4. & 4'. M03 — the clockwise revolution is ON.

G86 Boring Cycle

G86 X ...Y ... Z ... R ... F ...

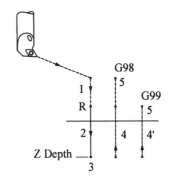

Figure 78

1. G00 —rapid traverse to R.
2. G01 —feed traverse to Z depth.
3. M05 —spindle revolution is stopped.
4. G00 —rapid traverse to the level plane assigned to function G98.
4'. G00 —rapid traverse to the R level plane for function G99.
5. M03 —the clockwise revolution is ON.

Part 4 Canned Cycle Functions

Example of Program No. 16

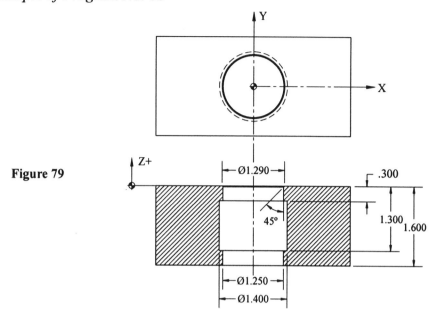

Figure 79

Machine: Machining Center	Program Number: O0016

Workpiece Zero X = <u>Upper Left Corner</u> Y = <u>Upper Left Corner</u> Z = <u>Top Surface</u>

Setup Description:

The material used is aluminum. Fasten the workpiece in the same way as in the previous example.

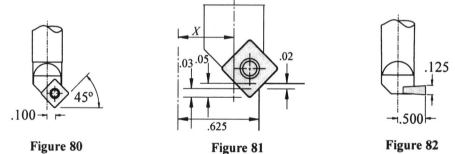

Figure 80 **Figure 81** **Figure 82**

$X = .625 - .3 - .05 = .545$

Tool #	Description	Offset	Comments
T01	#6 HSS CTR drill		SFM = 300 Feed =.002 inch per flute
T02	Drill 1.125 HSS		SFM = 300 Feed =.0035 inch per flute
T03	Carbide boring bar 1.25 diameter		SFM = 800 Feed =.001 inch per flute
T04	Carbide 45° chamfer cutter		SFM = 800 Feed =.001 inch per flute
T05	Carbide back boring bar 1.40 diameter .125 width insert		SFM = 800 Feed =.001 inch per flute

Part 4 Canned Cycle Functions

Program

O0016

N10G90G80G20G40G49

N15T01M06

(T01 CENTER DRILL FOR BORES)

N20G54G00X0Y0S2291M03

N25G43Z1.0H01M08

N30S2291M03

N35G81G98Z-.35F9.16R.1

N40G80Z1.0M09

N45G91G28Z0

N50M01

N55T02M06

(T02 DRILL 1.125 DIAMETER FOR BORES)

N60G90G80G40G49

N65G54G00X0Y0S1182M03T03

N70G43Z1.0H02M08

N75G73G98Q.3Z-2.0F8.27R.1

N80G80Z1.0M09

N85G91G28Z0

N90M01

N95T03M06

(T03 BORE 1.25 DIAMETER HOLE THROUGH)

N100G90G80G40G49

N105G54G00X0Y0S2444M03T04

N110G43Z1.0H03M08

N115G86G98Z-1.65F2.44R.1

N120G80Z5.M09

N125G91G28Z0

N130M01

N135T04M06

Part 4 Canned Cycle Functions

(T04 CHAMFER 45° X .02 ON 1.25 DIAMETER HOLE)

N140G90G80G40G49

N145G54G00X0Y0S3086M03T05

N150G43Z1.H04M08

N155S3086M03

N160G01Z-.05F50.0

N165X.545F3.08

N170G03I-.545

N175G01X0

N180G00Z1.0M09

N185G91G28Z0

N190M01

N195T05M06

(T05 BACK BORE 1.400 DIAMETER UNDERCUT)

N200G90G80G40G49

N205G54G00X0Y0S2182M03

N210G43Z1.0H05M08

N215S2182M03

N220G01Z-.425F50.0

N225G91G41Y.7F6.5D50

(D50 = .500)

N230G03J-.7F2.18

N235M98P0002L14

N240G01G40Y-.7F50.0

N245G90G00Z5.M09

N250G91G28Z0M05

N255G28X0Y0

N260M30

This subprogram is repeated 14 times to attain the full depth for the 1.400 diameter bore. A better method would be to use G87.

Part 4 Canned Cycle Functions

Subprogram

O0002

N10G01Z-.1F2.0

N15G03J-.7F3.0

N20M99

G87 Boring Cycle

Function G87 is used to bore that part of the hole or chamfer on the bottom of the hole (cutting is along the Z axis in the positive direction).

G87 X...Y...Z...R...Q...P...F...

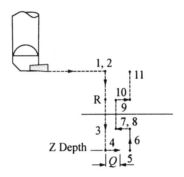

Figure 83

1. G00 — rapid traverse of oriented boring bar with a value of Q.
2. Non-programmed M19 — orientation of boring bar (same as for G76).
3. G00 — rapid traverse to Z (bottom of the hole).
4. G00 — rapid traverse of boring bar with a value of Q.
5. M03 — the clockwise rotations are ON.
6. G01 — work traverse to programmed Z value necessary to attain bore.
7. Non-programmed M19 — orientation of boring bar.
8. G00 — rapid traverse of boring bar with a value of Q.
9. G00 — rapid traverse to the R level plane.
10. G00 — rapid traverse with a value of Q.
11. M03 — the clockwise rotation is ON.

G88 Boring Cycle

G88 X...Y...Z...R...P...F...K...

Part 4 Canned Cycle Functions

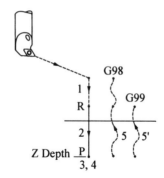

Figure 84

1. G00 — rapid traverse to R.
2. G01 — feed traverse to Z depth.
3. G04 — temporary interruption of feed for the time period P.

Dwell for the purpose of accurate removal of material from the bottom of the hole.

4. M05 — the revolution stop.
5., 5'. Manual or mechanical withdrawal of the tool.

After the revolutions are stopped, the tool may be removed from the holder while still at the bottom of the hole.

P. — time period of the temporary interruption is given in seconds.

G89 Boring Cycle

G89 X... Y... Z... R... P... F...

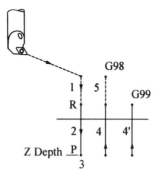

Figure 85

1. G00 — rapid traverse to R.
2. G01 — feed traverse to Z depth.
3. G04 — temporary interruption of feed.
4, 4' G01 — feed traverse to R.
5. G00 — rapid traverse to a level plane for function G98.

Part 4 Machining Center Example Programs

Examples of Programming Computer Numerically Controlled Machining Centers

Absolute (G90) Or Incremental (G91) Programming Comparison

In the absolute system, all dimensions are relative to the origin of the workpiece coordinate system.

In the incremental system, every measurement refers to the previously dimensioned position. Incremental dimensions are distances between the adjacent points (from the current tool location).

Example of Program No. 17

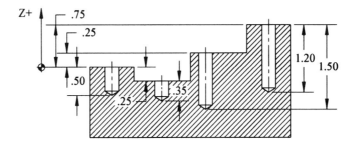

Figure 86

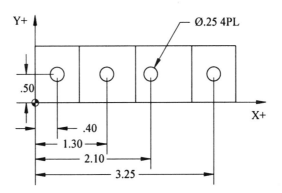

Machine: Machining Center	Program Number: O0017

Workpiece Zero X = <u>Lower Left Corner</u> Y = <u>Lower Left Corner</u> Z = <u>Top Surface</u>

Setup Description:

Material used is steel.

Tool one is mounted in the spindle prior to beginning.

Tool #	Description	Offset	Comments
T01	Tool is drill ¼ HSS.		SFM = 70 Feed = .0035 inch per flute

Part 4 Machining Center Example Programs

Selection Of Coordinate System (G92)

G92X(A)Y(B)

For easier and faster programming, each new program is assigned a new coordinate system to which all positioning is referred. In reality, zero of the new workpiece coordinate system is dimensioned from machine zero.

The following is a program employing an absolute coordinate system.

Program

O0017

N10G90G80G20G40G49

N15G92X10.0Y-7.0Z0S1070M03

N20G00X.4Y.5

N25G43Z1.0H01M08

N30G81G98Z-.5F7.48R.1

N35X1.3Z-.6R-.15

N40X2.1Z-.75R.35

N45X3.25Z-.45R.85

N50G80Z1.0M09

N55G91G28Z0M05

N60G28X0Y0

N65M30

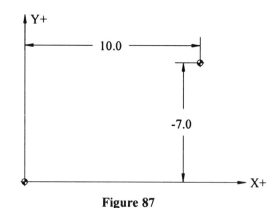

Figure 87

The following is an example of the same program as above employing an incremental coordinate system.

O0017

N10G90G80G20G40G49

N15G92X10.0Y-7.0Z0S1070M03

N20X.4Y.5

N25G43Z1.0H01M08

N30G91G81G98Z-.6F6.4R-.9

N35Z-.45R-1.15

N40X.9Z-1.1R-.65

N45X.8Z-1.3R-.15

Part 4   Machining Center Example Programs

N50G80X1.15Z5.0M09

N55G28Z0M05

N60G28Y0

N65X0

N70M30

Notes: Sign (+ or -) is determined with respect to the axis of the coordinate system.

Function G90 or G91 must be entered at the beginning of the program, where it remains valid until replaced by function G91 or G90.

Complex Program Example 1

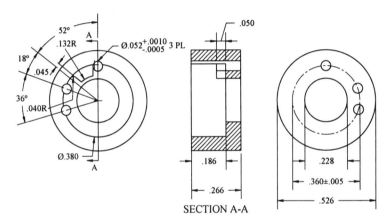

Figure 88

To machine the part represented by the preceding drawing, first use a lathe to prepare the rings dimensioned, as follows:

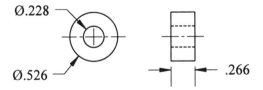

Figure 89

Consecutive operations will be performed on the vertical machining center. This part will be machined in two operations. First, three holes with a diameter of .052 will be drilled, and, second, the inside of the workpiece will be milled. Machining speeds, simplicity of fastening, and maximal accuracy are the decisive factors that determine the selection of the number of operations made and the sequence they follow.

The programs consist of two operations with tools given in the following set-up sheets:

Part 4 Machining Center Example Programs

The program for the first operation is specified by O0018. The program for the second operation is specified by O 0019.

Machine: Machining Center	Program Number: O0018, **Operation 1**

Workpiece Zero

G54 X = Upper Left Seat CL Y = Upper Left Seat CL Z = Top of Part Surface

G55 X = Lower Left Seat CL Y = Lower Left Seat CL Z = Top of Part Surface

Setup Description:

Material used is bronze.

Tool #	Description	Offset	Comments
T01	#0 HSS Center drill.	H01	
T02	$\frac{1}{64}$, HSS Drill	H02	
T03	.052 Diameter Reamer	H03	

Machine: Machining Center	Program Number: O0019, **Operation 2**

Workpiece Zero

G54 X = Upper Left Seat CL Y = Upper Left Seat CL Z = Top of Part Surface

G55 X = Lower Left Seat CL Y = Lower Left Seat CL Z = Top of Part Surface

Setup Description:

Material used is bronze.

Tool #	Description	Offset	Comments
T04	.125 Diameter HSS Roughing End Mill	H04	SFM = 70 Feed =.002 inch per flute
T05	.0781 Diameter HSS 4-Flute Finishing End Mill	H05	SFM = 70 Feed =.002 inch per flute

Fixture Preparation

Prepare the work holding fixture for the machined parts from a piece of rectangular steel, dimensioned as follows: 2.5 x 12.0 x .50 thick.

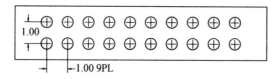

Figure 90

The fixture is fastened to the mill table with four bolts. The longer side of the part is milled with an end mill in order to establish parallelism with the part seats, which will later be machined. Additionally, this machined side will serve as a guide, in case another setup of the fixture is required. Afterwards, the 20 seats in the steel plate, in which the parts will be fastened, are machined.

Part 4 Machining Center Example Programs

The seats have the following dimensions: diameter .528, depth .100. Between the seats, drill and then tap holes for part clamping screws. For this example, the work-holding fixture has been prepared earlier.

Complex Program Example 1 Setup Description

The fixture is placed on the machining center table and is positioned so that the milled side is parallel to the X axis. To confirm this position, use a dial indicator to dial-in the setup of the fixture. As soon as parallelism with the X axis is verified, the fixture is clamped to the table with machine screws. The alignment of the fixture is checked again, after clamping, to assure that the fixture did not move during the fastening. Next, the values of X and Y for the functions G54 and G55 are determined.

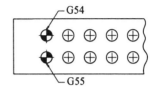

Figure 91

Procedure for Measuring Work Offsets

Detailed descriptions for set-up procedures are given in Part II of this book but a brief description is given here as well, to clarify application to this example. Remember that the procedures described are specific to the type of control described in this text (Fanuc 16-18). Other control types use similar procedures, check the specific manufacturer operation manuals for their procedures.

First of all, zero the machine with respect to the X and Y axes. By pressing the POSITION button (POS function key), the values of the X, Y, and Z will appear on the display.

By pressing the POSITION function button, the absolute coordinate display can be found. Press the soft key labeled "ABS" to display the absolute coordinates of X, Y, and Z. This page shows the values of the coordinates of the axes positioned at machine zero.

After machine zero is established, check to verify that the absolute display values shown are zero. If not, then press button X and then press the ORIGIN soft key (X, Y ORIGIN). Now position the machine in order to determine the centers of the holes, fasten an indicator in a drill chuck placed in the spindle. Use the HANDLE mode to move the axes to find the center of the first left hole positioned in the upper row of holes. The values of X and Y are read from the absolute display and entered under the offset values of the coordinate system, of which position 01 corresponds to the function G54. The position of the left hole in the lower row is determined in a similar manner and the values of the coordinates corresponding to function G55 are entered under 02, in the coordinate system register. Once workpiece zero is set, place the previously prepared bronze bushings in the fixture seats and fasten them with machine screws.

Procedure for Measuring Tool Length Offsets

Place the center drill in a drill chuck. Set the mode to MDI, key in T01M06; press INPUT and then CYCLE START to install the tool into the spindle. Next, change back to the HANDLE mode. By using the function button OFFSET/SETTING and the OFFSET soft key the tool-length-offsets page is found. Check the LED for the Z axis, which, for this axis, corresponds to machine zero. If the LED is on, then for the Z axis, the

Part 4   Machining Center Example Programs

machine is at zero. The values of X, Y, or Z are shown on the bottom of the screen. Check whether the value of Z is equivalent to zero. If it is not, then press the buttons ORIGIN and then Z to zero it. To determine the tool length offset, position the tool to touch the surface of the machined part with the tool tip. Lower the tool manually using the pulse generator (Handle) with caution, reducing the traverse speed as the tool approaches the surface. If a touch control edge finder is used (touch sensor), then as soon as the tool tip touches the surface, the LED is lit. (Touch control provides a very accurate measurement of the tool length offset.) When the control LED lights, the value of Z must be read from the readout and entered into the tool-length-offset register under 01, which corresponds to tool 1 (T01). However, if no touch sensor is used another common method of determining the tool length offset is given as follows.

Place a piece of paper with a given thickness between the tool tip and the surface of a machined part. While approaching the surface of the machined part with the tool tip, move the piece of paper back and forth. As soon as the first resistance to the paper is felt, stop the traverse and withdraw the piece of paper. Next, check the displayed readout and then lower the tool a distance equivalent to the thickness of the paper.

Two methods of entering the value of Z to the tool-length-offset tables are listed below:

The first method (directly): As soon as the tool tip touches the surface of a machined part, press the button Z. If the letter Z is flashing on the screen, then it is possible for the machine to transfer the value of Z directly to the offset register. By pressing the INP.C. soft key, the value of Z is transferred to the tool-length-offset register.

The second method (indirectly): The value of Z is read from the displayed readout and noted. Using the cursor move keys, the cursor is positioned at the desired offset. The value noted previously is keyed in and the INPUT button pressed. The value of Z is transferred to a previously reserved position 01 of the tool length offset (a center drill). A similar principle is applied to determine the tool length offsets for the remaining tools (a drill and a reamer), as well as for the tools in the second operation. The tool is removed from the spindle and then transferred to the tool magazine. The machine is set at zero in the Z axis prior to each tool measurement.

The following is the main program for the first operation.

Example of Programs No. 18 and No. 19

```
O0018
N10G90G80G20G40G49
N15T01M06
N20G54G00X0Y.1787S4000M03T02
N25G43Z1.H01M08
N30G81G98Z-.06R.05F4.0
N35M98P2
N40G90G00G55X0Y.1787
```

Part 4 Machining Center Example Programs

N45G81G98Z-.055R.05F4.0

N50M98P2

N55G90G00G80Z5.M09

N60G91G28Z0

N65G28X0Y0

N70M01

N75T02M06

N80G90G80G40G49

N85G00G54X0Y.1787S4000M03T03

N90G43Z1.H02M08

N95G83G98Z-.35R.05Q.05 F4.0

N100M98P2

N105G90G00G55X0Y.1787

N110G83G98Z-.35R.05Q.05F4.0

N115M98P2

N120G80G00Z1.0M09

N125G91G28Z0

N130G28X0Y0

N135M01

N140T03M06

N145G90G80G40G49

N150G54G00X0Y.1787S4000M03

N155G43Z1.H03M08

N160G85G98Z-.31R.05F6.0

N165M98P2

N170G80G90G00G55X0Y.1787

N175G85G98Z-.31R.05F6.0

N180M98P2

N185G80G00Z1.0M09

N190G91G28Z0

N195G28X0Y0

Part 4   Machining Center Example Programs

N200M05

N205M30

The following is the subprogram O0002 for the first operation.

O0002

N2X-.1679Y.0611

N3X-.1718Y-.0493

N4X1.Y.1787

N5X.8321Y.0611

N6X.8282Y-.0493

N7X2.Y01787

N8X1.8321Y.0611

N9X1.8282Y-.0493

N10X3.0Y.1787

N11X2.8321Y.0611

N12X2.8282Y-.0493

N13X4.0Y.1787

N14X3.8321Y.0611

N15X3.8382Y-.0493

N16X5.0Y.1787

N17X4.8321Y.0611

N18X4.8282Y-.0493

N19X6.0Y.1787

N20X5.8321Y.0611

N21X5.8282Y-.0493

N22X7.0Y.1787

N23X6.8321Y.0611

N24X6.8321Y-.0493

N25X8.0Y.1787

N26X7.8321Y.0611

N27X7.8282Y-.0493

N28X9.0Y.1787

Part 4 Machining Center Example Programs

N29X8.8321Y.0611

N30X8.8282Y-.0493

N31M99

The following is the main program for the second operation.

O0019

N10G90G80G20G40G49

N15T04M06

N20G54G00X0Y0S3000M03T05

N25G43Z1.H04M08

N30M98P6L10

N35G90G00G55X0Y0

N40M98P6L10

N45G00G90Z1.0M09

N50G91G28X0Y0Z0

N55M01

N60T05M06

N65G90G80G40G49

N70G00G54X0Y0S2300M03T04

N75G43Z1.H05M08

N80M98P7L10

N85G00G90G55X0Y0

N90M98P7L10

N95G00G90Z1.0M09

N100G91G28X0Y0Z0

N105M01

N110M30

Subprogram O0006 is for the second operation.

O0006

N1G90G00Z0

N2G91G01Z-.182F4.0

N3G42X-.1225D01

Part 4   Machining Center Example Programs

N4G02X.0138Y.0564I.1225

N5G01X.0412Y-.0369

N6G02X.088Y.0393I.0615J-.0195

N7G01Y.0608

N8G02X.1352Y-.0632I-.0265J-.0195

N9G01Z.05

N10G02X.1352Y.0632I.1087J-.1196

N11G01G40X-.0265Y-.1196

N12G90G00Z1.0

N13G91X1.0

N14M99

Subprogram O0007 is for the second operation.

O0007

N1G90G00Z0

N2G91G01Z-.184F2.0

N3G01G42X-.151D02

N4G02X.0358Y.0976I.151F3.0

N5G01X.0458Y-.0357F1.0

N6G02X.0674Y.0311I.0694J-.0619F2.0

N7G01Y.058 F.5

N8G02X-.1132Y-.0534I.002J-.151F2.0

N9G01Z.05F10.0

N10G02X.1132Y.0534I.1152J-.0976F2.0

N11G01G40X.002Y-.151

N12G90G00Z1.0

N13G91X1.0

N14M99

The following explanations are for some blocks of program O0018, operation one.

N10G90G80G20G40G49

Block N10 is the safety block and contains the functions G90, G80, G20, G40, and

Part 4 Machining Center Example Programs

G49 where G90 refers to programming in an absolute coordinate system (all dimensions correspond to program zero), G80 is entered in the beginning of the program to assure that all canned cycle functions are cancelled, G20 initiates the inch measurement system, G40 assures that all cutter diameter compensation functions are cancelled and G49 cancels all tool length compensations.

N15T01M06

In this block, Tool T01 is called from the magazine and is inserted into the spindle.

N20G54G00X0Y.1787S4000M03T02

In this block, the control reads the coordinates of the program zero (defined in G54) from the coordinate system offset registers and then moves at rapid traverse, with respect to that zero, the distance whose value is given by X and Y (in this case, X0 and Y.1787). This block also defines the spindle rotation of 4000 (S4000) revolutions per minute (RPM) in the clockwise direction (M03) and calls tool number 2 (T02), into the ready position in the magazine.

N25G43Z1.H01M08

In this block, the control reads the value of the positive tool length offset (G43) for the center drill, as is registered in offset #1 and called in the program by H01 and positions the tool at rapid traverse along the Z axis corresponding to the values of one inch above the part. Miscellaneous function, M08 activates the flood coolant flow.

N30G81G98Z-.060R.05F4.0

Functions G81 and G98 refer to the canned drilling cycle to a depth of -.060. At the completion of cycle the drill rapidly traverses to the reference plane identified by the letter R, of .050 inch above the material, and, for positioning to the remaining holes, the drilling begins from this point and feeds down to a depth of Z-.060 with a feed rate of F4. inches-per-minute. After completing all of the spot drilling, the center drill withdrawal occurs with a rapid motion (G00) to the Z value, which is given in block N25, Z1.

N40M98P2

Function M98 calls up subprogram O0002 from the controller memory. From this point on, execution of subprogram O0002 will be in progress until its completion.

The following explanations are for a few blocks of subprogram O0002 called from the main program O0018, operation one.

N2X-0.1679Y0.0611

In block N2, the center drill performs the spotting of the second hole with the assigned coordinates X and Y.

N3X-0.1718Y-0.0493

In this block, the third hole is drilled in the first bushing.

N4X1.0Y0.1787

The center drill is displaced along a horizontal straight line (along the X axis) by the value X1.0 over to the second bushing and the first hole is spot drilled. Similar moves are performed in consecutive blocks and consecutive holes are spot drilled. In block N30 of the subprogram, spot drilling of the holes in the 10th bushing is completed.

Part 4 Machining Center Example Programs

N31M99

Block N31 ends the subprogram, and function M99 commands the return to the main program O0018, to block N40.

The explanations that follow return to the main program O0018, at block N40:

In block N40 of the main program, the new program zero G55 is assumed in the center of the first bushing in the lower row. In block N45, the drilling command is repeated and in block N50, the function M98 P2 commands the repeated execution of the subprogram O0002. Executing the subprogram causes the machine to spot-drill holes in the 9 remaining bushings (the lower row). When the work is completed, there is a return to block N55 of the main program, a deactivation of coolant flow M09, and, simultaneously, a withdrawal of the spotting drill to Z1.0 along the Z axis at rapid traverse. In block N60, the tool is moved to the automatic reference point return value G28 with respect to the Z axis and then in block N65 the X and Y axes are moved as well. Function M01 is entered in block N70. Work is interrupted in this block, but only if the OPTIONAL STOP button is ON. In block N75, tool two, (T02), is called into position in the magazine and the tool change takes place in M06.

N80G90G80G40G49

This block contains the safety block information required after each tool change.

N85G00G54X0Y.1787S4000M03T03

Function G54 refers to the return from the previous coordinate system, i.e., the functions G54 to G55 (program zero is located in the center of the first bushing in the upper row). The tool then rapidly traverses (G00) to the position above the first hole, with coordinates X0 and Y.1787.

N90G43Z1.H02M08

The negative value of the tool length offset from the offset register for T02 is read by the control. The tool rapidly traverses along the Z axis to a position one inch above the material. The coolant flow is activated by the function M08. Block N90 is similar to block N25. They differ only by the value of the tool length offset H02, because tool T02 is used.

Comparing blocks N15 through N70 with blocks N75 through N135 and N140 through N200, notice the similarities:

N15 through N70 — work of the first tool.

N75 through N135 — work of the second tool.

N140 through N200 — work of the third tool.

In block N205, notice the function M30, it stops the execution of the program and commands the program to return to its beginning. After completion of block N205, the bushings are replaced in the fixture by new ones and then by pressing the CYCLE START button, machining of the next 20 pieces is initiated.

Part 4 Machining Center Example Programs

The Second Operation

In the second operation, the drilled bushings are milled using two end mills, first a rough pass with a tool diameter of .125, and, second, a finish pass with a tool diameter of $5/64$ in. The bushings are placed in a fixture similar to the one in the previous operation. The only difference in the second fixture, is that each seat has a pin forced into the bottom with a diameter of .0515 onto which the previously reamed hole in the bushing is placed. Such positioning of the bushings assures the proper position of the hole and prevents rotation of the bushing in the seat during milling.

Main Program Number O0019

The following explanations are for some blocks of program O0019 operation two.

Subprogram O0006 is for the first tool and subprogram O0007 is for the second tool.

N10G90G80G20G40G49

This is a safety block where G90 sets the program in the absolute coordinate system. G80 cancels all canned cycle functions. G20 initiates the inch measurement system. G40 cancels all radius compensation functions. G49 cancels all tool length compensation functions.

N15T04M06

This block commands the preparation of the tool T04 in the magazine and changes it into the spindle.

N20G54G00X0Y0S3000M03T05

Workpiece zero is entered by function G54; there is a rapid traverse of the tool to the position of that zero. The spindle is started in the clockwise direction, M03, at a speed, S, of 3000 RPM and tool number 5 (T05) is moved to the ready position in the magazine.

N25G43Z1.H04M08

In this block, the machine is commanded to read the tool length offset for T04 from the offset register, while the tool positions at rapid traverse to 1 inch above the material along the Z axis. The flood coolant in activated by M08.

N30M98P6L10

The subprogram, O0006 is called up and executed nine times (for the 10 bushings). At this point, the tool is controlled by the subprogram. In order to simplify the program notation, the position of eight characteristics points on the drawing of the part is given below.

Coordinates of the individual points are

P_0: X0, Y0 P_1: X − .1255, Y0 P_2: X − .1087, Y + .0564
P_3: X − .0615, Y + .0195 P_4: X + .0265, Y + .0588 P_5: X + .0265, Y + .1196
P_6: X − .1087, Y + .0564 P_7: X + .0265, Y + .1169 P_8: X0, Y0

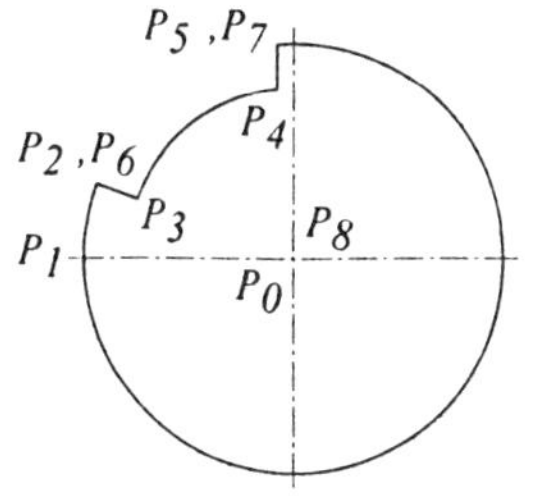

Figure 92

Part 4 Machining Center Example Programs

The following explanations are for a few blocks of subprogram O0006 called from the main program O0019, operation two.

Subprogram O0006

N1G90G00Z0

The block refers to programming in the absolute coordinate system and traverses along the Z axis, at rapid, to position 0.

N2G91G01Z-.182F4

Converts the coordinate system from the absolute to incremental.

The tool traverses in a linear motion to Z-.182 with a feed rate of 4 IPM.

N3G01G42X-.1225D01

The tool traverses with a previously assigned feed rate to a position of X-.1225, with a simultaneous reading of the radius offset (D01), from the offset register. (In this case, the value of the offset is equivalent to zero, because the traverse of the end mill is defined so that the distance from the center of the end mill to the theoretical dimension is exactly equivalent to the tool radius.) In addition, 0.005 inch of material remains for the finishing operation, in which the finishing end mill is used.

N4G02X.0138Y.0564I.1225

This feeds the tool along a circular path, in a clockwise direction, with a radius vector value of I.1225 (the center of the end mill is located in P_2).

N5G01X.0412Y-.0369

The tool traverses along a straight line, specified by the values of X and Y (the center of the end mill is located in P_3).

N6G02X.088 Y.0393I.0615J-.0195

Traverse along a circular path, in a clockwise direction, specified by the values of X and Y (the center of the end mill is located in P_4). Values in I and J denote the incremental distance to the arc center point.

N7G01Y.0608

Traverse along, straight line, specified by the value of Y (the center of the end mill is located in P_5).

N8G02X.1352Y-.0632I-.0265J-.0195

Traverse along the circular path to P_6.

N9G01Z.050

The end mill is retracted from the part, at feed rate, by a value of .050.

N10G02X.1352Y.0632I.1087J-.1196

Traverse along a the circular path to P_7

Part 4 Machining Center Example Programs

N11G01G40X-.0265Y-.1196

The active function G42 is cancelled by G40, and the tool is displaced along a straight line to P_8

N12G90G00Z1.0

The tool is positioned, at rapid traverse, 1 inch above the material (use G90 to assure that the tool position is above the material).

N13G91X1.0

A rapid traverse is made along the X axis by a value of 1 in (toward the center of the second bushing).

N14M99

This block commands a return to the main program O0019. However, because function M98 L10 was entered previously in the main program, the subprogram will be executed nine more times (nine bushings will be machined in exactly the same manner). After the machining of the last bushing in the upper row is completed, control is returned to block N35 of the main program.

N35G90G00G55X0Y0

In this block, the new workpiece zero G55 is introduced with its assigned coordinates given in the offset register. G00 refers to a rapid traverse to the new position of the new workpiece zero. (Position of the new workpiece zero is set to the center of the first bushing of the lower row.)

N40M98P6L10

Function M98 calls up subprogram O0006 and executes it 10 times (subprogram O0006 was described earlier). After the 10 executions of the subprogram, control is returned to block N45 of the main program for the second operation.

N45G00G90Z1.0M09

A rapid traverse is made to a position 1 inch above the surface of the material based on the absolute coordinate system, and the deactivation of the coolant flow (M09) occurs.

N50G91G28X0Y0Z0

Returns the machine zero in the X, Y and Z axes.

N55M01

Provides an additional interruption of the program (only if the OPTIONAL STOP button is ON). The purpose of this interruption is to check the dimensions and condition of the tools.

N60T05M06

This is tool change; it transfers the tool T04 to the tool magazine and positions the tool T05 in the spindle of the machine.

N65G90G80G40G49

This is a safety block.

N70G00G54X0Y0S2300M03T04

Part 4   Machining Center Example Programs

A new program zero is introduced; the tool rapidly travels to the position of the new zero and clockwise rotation is initiated at 2300 RPM and tool number 4 is readied in the magazine for change.

N75G43Z1.H05M08

The tool length offset H05, obtained from the offset register, is entered, allowing the traverse of the tool to a position 1 inch above the surface of the material. Flood coolant flow is also activated.

N80M98P7L10

The subprogram O0007 is called up and then executed 10 times.

P_0: X0, Y0 P_1: X−.151, Y0 P_2: X−.1152, Y+.0976
P_3: X−.0694, Y+.0619 P_4: X−.002, Y−.093 P_5: X−.002, Y+.151
P_6: X−.1152, Y+.0976 P_7: X−.002, Y+.151 P_8: X0, Y0

The previous set of coordinates illustrate the position of the center of the end mill. Points P_0 through P_8 correspond to the characteristic path points of the end mill.

N1G90G00Z0

Performs the rapid traverse of the end mill toward the surface of the material.

N2G91G01Z-.184F2.0

Establishes measurement in the incremental system.
The work traverses along the Z axis with a feed rate of 2 IPM.

N3G01G42X.151D02

The radius offset is read, and there is a traverse along the X axis to the value X-.151 (point P_1).

N4G02X.0358Y.0976I.151F3.0

The traverse is along a circular path in a clockwise direction to point P_2

N5G01X.0458Y-.0357F1.0

The traverse is along a straight line to point P_3

N6G02X.0674Y.0311I.0694J-.0619F2.0

The traverse is along a circle in a clockwise direction to point P_4.

N7G01Y.058 F5.0

The traverse is along a straight line to point P_5.

N8G02X-.1132Y-.0534I.002J-.151F2.0

The traverse is along a circular path to point P_6.

N9G01Z.05F10.0

The tool is raised along the Z axis by a distance of .05 in.

Part 4 Machining Center Example Programs

N10G02X.1132Y.0534I.1152J-.0976F2.0

The traverse is along a circular path to point P_7.

N11G01G40X.002Y-.151

The tool radius offset is cancelled and the traverse is along a straight line to P_8, the center of the bushing.

N12G90G00Z1.0

A rapid tool traverse is made to a position 1 inch above the surface of the material, defined in an absolute coordinate system.

N13G91X1.0

A rapid traverse is along the X axis for 1 inch (to the position of the second bushing).

N14M99

This block commands a return to the main program. Since command L10 was included in the main program, the subprogram will be executed nine more times before there is a return to block N85 of the main program.

N85G00G90G55X0Y0

A new coordinate system is assigned to function G55, and the tool moves at rapid traverse to the zero position of the new system.

N90M98P7L10

The subprogram O0007 is called up, executes the subprogram 10 times, and then returns to N95.

N95G00G90Z1.0M09

A rapid traverse to a position 1 inch above the surface of the material defined in the absolute coordinate system and the coolant will be turned off.

N100G91G28X0Y0Z0

There is a return to machine zero for the Y and Z axes.

N105M01

There is an additional interruption of the machine work, providing the OPTIONAL STOP switch/button is ON.

N110M30

This is the end of the program and control returns to the beginning. Completion of N110 results in a machine stop. The finished bushings must be removed and replace them with unfinished ones. Press the CYCLE START button and the work cycle is initiated once again.

Part 4   Machining Center Example Programs

Example Illustrating the Application of the Mirror Image

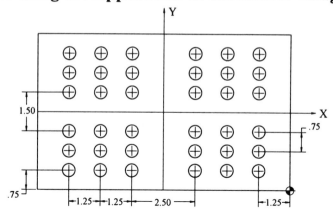

Figure 94

In the following example, use of the mirror image function is illustrated. This function makes it possible to duplicate patterns by projecting a mirror image of the pattern in the X or Y axis directions. The main purpose for this is to shorten and simplify the program.

M21 = the mirror image in the direction of the X axis.

M22 = the mirror image in the direction of the Y axis.

M23 = the cancellation of the mirror image.

Example of Program No. 20

```
O0020
N10G90G80G20G40G49M23
N15G00G90G54X-1.25Y.75S1000M03
N20G43Z1.0H01M08
N25G81G98Z-.35R.1F6.0
N30M98P2
N35G00G80X0
N40M21
N45G00X-1.25Y.75
N50G81G98Z-.35F6.R.1
N55P2M98
N60G00G80X0Y0
N65M23
N70M22
N75G00X-1.25Y.75
```

Part 4 Machining Center Example Programs

N80G81G98Z-.35F6.R.1

N85M98P2

N90G80G00X0Y0

N95M21

N100G00X-1.25Y.75

N105G81G98Z-.35F6.R.1

N110M98P2

N115G80Z1.0M09

N120M23

N125G91G28X0Y0Z0

N130M30

Subprogram forExample 00020

   O0002

   N1X-2.5

   N2X-3.75

   N3Y1.5

   N4X-2.5

   N5X-1.25

   N6Y2.25

   N7X-2.5

   N8X-3.75

   N9M99

Example of a Program With the Application of a Rotation With Respect To the X Axis

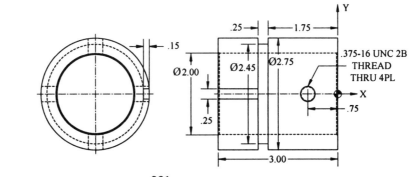

Figure 95

Part 4   Machining Center Example Programs

Occasionally, vertical milling machines are used to perform operations on cylindrical or other workpieces that require the use of a rotational axis. In the following case, there are four drilled and tapped holes, and a slot that must be machined by using this method.

Machine: Machining Center	Program Number: O0021

Workpiece Zero

G54 X = <u>Right End</u> Y = <u>Center Line</u> Z = <u>Top of Part Surface</u>

Setup Description:

Material used is Steel.

Mount the part in a chuck attached to an indexing head or 4th axis.

Tool #	Description	Offset	Comments
T01	#3 HSS Center Drill		SFM = 95 Feed =.002 inch per flute
T02	.3125 Diameter HSS Drill		SFM = 95 Feed =.0035 inch per flute
T03	Tap $\frac{3}{8}$ - 16		SFM = 15
T04	.250 HSS 4-flute End Mill		SFM = 95 Feed =.002 inch per flute

Program

O0021

N10G90G80G20G40G49

N15T01M06

N20G00G54X-.75Y0A0S1520M03

N25G43Z1.H01M08

N30G98G81Z-.375R.1F6.0

N35M98P2

N40G80Z1.0M09

N45G91G28X0Y0Z0

N50M01

N55T02M06

N60G90G80G40G49

N65G54G00X-.75Y0A0S1152M03

N70G43Z1.H02M08

N75G98G81Z-.450R.1F8.0

N80M98P2

Part 4   Machining Center Example Programs

N85G80Z1.0M09

N90G91G28X0Y0Z0

N95M01

N100T03M06

N105G90G80G40G49

N110G00G54X-.75S120M03

N115G43Z1.H03M08

N120G98G84Z-.55R.2F7.5

N125M98P2

N130G80Z5.M09

N135G91G28X0Y0Z0

N140M01

N145T04M06

N150G90G80G40G49

N155G54G00X-3.3Y0A0

N160G43Z1.H04M08

N165S1520M03

N170G01Z-.15F50.0

N175X-1.875F6.0

N180A360

N185G00Z1.0M09

N190G91G28X0Y0Z0A0

N200M30

If, in a given block, there is a rotation during the work traverse, then, the feed is defined in degrees per revolution.

Subprogram for example program O0021.

    O0002

    N1A90

    N2A180

    N3A270

    N4M99

Part 4 Machining Center Example Programs

Example Of Programming A Horizontal Machining Center

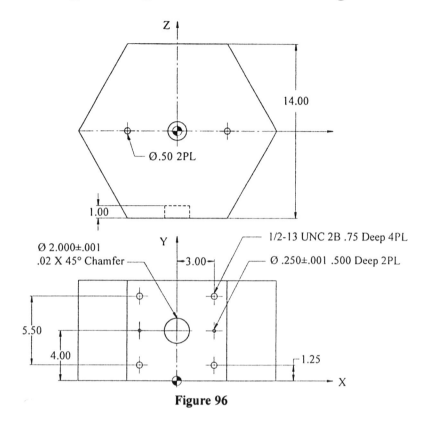

Figure 96

Thus far, in this part of the text, only vertical machining center programming has been described. In this example, some of the programming techniques covered earlier will be applied to a horizontal machining center. With the horizontal machine, the Z axis is horizontal. Another feature that will be useful on the illustrated part is its indexing capability.

Example of Program No. 22

The material is a hexagonal plate with the following dimensions: 8 x 14 inch. There are also two positioning holes in the plate having .500-in diameter.

The workpiece is placed on the fixture that has two positioning pins that are .500 inch in diameter. The workpiece is held down by a holding mechanism and a cap screw that is placed in the center of a previously made hole of 1.5 inch. To simplify programming, assume that the rotation axis of the table is aligned with the symmetry axis of the object.

Part 4 Machining Center Example Programs

| Machine: Machining Center | Program Number: O0022 |

Workpiece Zero

G54 X = Center Line Y = Bottom of Part Surface Z = Center Line

Setup Description: See above

Tool #	Description	Offset	Comments
T01	#3 HSS Center Drill		
T02	.125 Diameter HSS Drill		
T03	1.0 Diameter 4-flute HSS End-Mill		
T04	$\frac{27}{64}$ (.4219) Diameter HSS Drill		
T05	$\frac{1}{2}$ – 13 UNC Tap		
T06	$\frac{15}{64}$ (.2344) Diameter HSS Drill		
T07	.240 HSS 4-flute End-Mill		
T08	.250 HSS Reamer		
T09	Boring Bar for 2.00 Diameter		
T10	45° Chamfer Cutter		

Program

O0022

N1G90G80G20G40G17

N2M11T01

N3M06

N4G90G00G54X0Y4.0B0S600M03

N5M10

N6G43Z8.5M08H01

N7P0002L6M98

N8G80Z10.0M09

N9M11

N10G91G28X0Y0Z0B0

N11M01

N12T02

N13M06

N14G90G80G20G40G17

285

Part 4 Machining Center Example Programs

N15G0G54X0Y4.0B0S244M03

N16M10

N17G43Z8.5M08H02

N18P0004L6M98

N19G80Z10.0M09

N20M11

N21G91G28X0Y0Z0B0

N22M01

N23T03

N24M06

N25G90G80G20G40G17

N26G00G54X0Y4.0B0S305M03

N27M10

N28G43Z8.5M08H03

N29P0005L6M98

N30G80Z10.0M09

N31M11

N32G91G28X0Y0Z0B0

N33M01

N34T04

N35M06

N36G90G80G20G40G17

N37G00G54X-3.0Y6.75B0S726

N38M10

N39G43Z8.5M08H04

N40P0006L6M98

N41G80Z10.0M09

N42M11

N43G91G28X0Y0Z0B0

N44M01

N45T05

Part 4 Machining Center Example Programs

N46M06

N47G90G80G20G40G17

N48G00G54X-.30Y6.75B0S100M03

N49M10

N50G43Z8.5M08H05

N51P0007L6M98

N52G80Z10.0M09

N53M11

N54G91G28X0Y0Z0B0

N55M01

N56T06

N57M06

N58G90G80G20G40G17

N59G00G54X3.0Y4.0B0S1300M03

N60M10

N61G43Z8.5M08H06

N62P0008L6M98

N63G80Z10.0M09

N64M11

N65G91G28X0Y0Z0B0

N66M01

N67T07

N68M06

N70G90G80G20G40G17

N71G00G54X-3.0Y4.0B0S1200M03

N72M10

N73G43Z8.5M08H07

N74P0009L6M98

N75180Z10.0M09

N76M11

N77G91G28X0Y0Z0B0

Part 4   Machining Center Example Programs

N78M01

N79T08

N80M06

N81G90G80G20G40G17

N82G00G54X-3.0Y4.0B0S800M03

N83M10

N84G43Z8.5M08H08

N85P0010L6M98

N86G80Z10.0M09

N87M11

N88G91G28X0Y0Z0B0

N89M01

N90T09

N91M06

N92G90G80G20G40G17

N93G00G54X0Y4.0B0S500M03

N94G43Z8.5M08H09

N95G76G98Q.01Z6.0F2.0R7.2

N96B60

N97B120

N98B180

N99B240

N100B300

N101G80Z10.0M09

N102G91G28X0Y0Z0B0

N103M01

N104T10

N105M06

N106G90G80G20G40G17

N107G00G54X0Y4.0S500M03

N108G43Z8.5M08H10

N109G82G98Z6.9F2.5R7.1

Part 4 Machining Center Example Programs

N110B60

N111B120

N112B180

N113B240

N114B300

N115G80Z10.0M09

N116G91G28X0Y0Z0B0M05

N117M10

N118M06

N119M30

Notes: There are two types of horizontal milling machines.

In milling machines equipped with an indexing table, rotation of the table only by 1° or more is allowed. When programming the changes of angular position of the table, as a rule, a decimal point is not used (but will do no harm).

Example:

B120, B180, etc.

Special functions are used to determine the direction of rotation of the table.

In a milling machine equipped with a four-axes table, rapid rotation or regular feed rotation can be applied. The table can be rotated by .001° or even more accurately. On these machines performance of all kinds of curved contours may be accomplished while machining the object. Plus (+) or minus (-) sign determined the direction of rotation. The table should not be locked if a rotation is to be performed. At the time tools 1 to 8 perform their operations, the table is locked by function M10. When the remaining two tools perform their operations, however, the table is unlocked by function M11. Whether the table should be locked or not is determined by the conditions of machining (light cutting/table unlocked or heavy cutting/table locked).

Subprograms for example O0022

O0002

N1G81G98Z6.5F2.5R7.1

N2X-3.0Y6.75

N3P0003M98

N4X-3.0Y4.0Z6.73

N5X3.0

N6G80X0 Y4.0M11

N7G91G00A60

N8G90M10

N9M99

Part 4 Machining Center Example Programs

O0003

N1X3.0

N2Y1.25

N3X-3.0

N4M99

O0004

N1G81G98Z6.0F1.5R7.1

N2G80M11

N3G91G00A60

N4G90M10

N5M99

O0005

N1G90G00Z7.1

N2G01Z6.0F3.0

N3G42Y5.0D50F2.5

N4G02J1.0

N5G01G40Y4.0

N6G00Z8.5M11

N7G91A60

N8G90M10

N9M99

O0006

N1G83G98Q.15Z5.8F4.5R7.1

N2P0003M98

N3G80X-3.0Y6.75M11

N4G91G00A60

N5G90M10

N6M99

Part 4 Machining Center Example Programs

O0007

N1G84G98Z6.0F7.2R7.3

N2P0003M98

N3G80X-3.0Y6.75M11

N4G91G00A60

N5G90M10

N6M99

O0008

N1G83G98Q.100Z6.25F6.5R7.1

N2X3.0

N3G80X-3.0Y4.0M11

N4G91G00A60

N5G90M10

N6M99

O0009

N1G81G98Z6.4F8.0R7.1

N2X3.0

N3G80X-3.0Y4.0M11

N4G91G00A60

N5G90M10

N6M99

O0010

N1G81G98Z6.47F4.8R7.1

N2X3.0

N3G80X-3.0Y4.0M11

N4G91G00A60

N5G90M10

N6M99

Part 4 Machining Center Example Programs
Complex Program Example 2

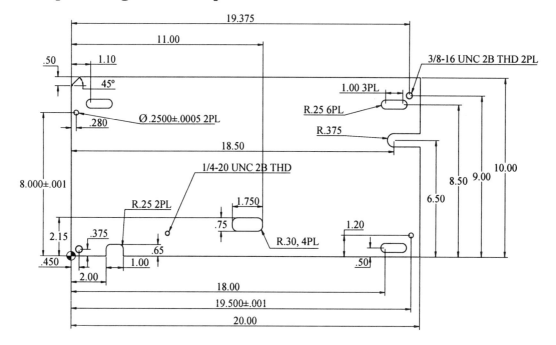

Figure 97

| Machine: Machining Center | Program Number: O0023 |

Workpiece Zero

G54 X = <u>Lower Left Corner</u> Y = <u>Lower Left Corner</u> Z = <u>Top Part Surface</u>

Setup Description:

The material is .25 thick steel.

Tool #	Description	Offset	Comments
T01	#3 HSS Center Drill		
T02	.3125 Diameter HSS Drill		
T03	3/8 – 16 UNC Tap		
T04	.201 Diameter HSS Drill		
T05	1/4 – 20 UNC Tap		
T06	.242 Diameter HSS Drill		
T07	.250 HSS Reamer		
T08	7/16 Diameter HSS Drill		
T09	7/16 Diameter Roughing End-Mill	D51	D51 Offset = .240
T10	7/16 Diameter Finishing End-Mill	D52	D52 Offset = .2187
T11	1.0 Diameter HSS End-Mill		

Part 4 Machining Center Example Programs

Example of Program No. 23

O0023
(T01 #3 CENTER DRILL)
N10G90G80G20G40G49
N15T01M06
N20G54G00X.45Y.375S611M03
N25G43Z1.0H01M08
N30G81G98Z-.375R.1F2.4
N35X19.375Y9.0
N40X19.500Y1.2Z-.26
N45X.28Y8.0
N50X5.5Y1.275Z-.25
N55X1.1Y8.5Z-.437
N60X18.0
N65Y.5
N70X10.5Y1.775
N75G80Z1.0M09
N80G91G28Z0
N85M01
(T02 5/16 DIAMETER TAP DRILL)
N90T02M06
N95G90G80G40G49
N100G54G00X.450Y.375S733M03
N105G43Z1.0H02M08
N110G81G98Z-.35R.1F4.3
N115X19.375Y9.0
N120G80Z1.0M09
N125G91G28Z0
N130M01
(T03 3/8-16 UNC 2B TAP)
N135T03M06

Part 4 Machining Center Example Programs

N140G90G80G40G49

N145G54G00X.450Y.375S150M03

N150G43Z1.0H03M08

N155G84G98Z-.45R.2F9.0

N160X19.375Y9.0

N165G80Z1.0M09

N170G91G28Z0

N175M01

(T04 .201 DIAMETER TAP DRILL)

N180T04M06

N185G90G80G40G49

N190G54G00X5.5Y1.275S1140M03

N195G43Z1.0H04M08

N200G81G98Z-.35R.1F6.8

N205G80Z1.0M09

N210G91G28Z0

N215M01

(T05 1/4-20 UNC 2B TAP)

N220T05M06

N225G90G80G40G49

N230G54G00X5.5Y1.275S220M03

N235G43Z1.0H05M08

N240G84G98Z-.42R.2F11.0

N245G80Z1.0M09

N250G91G28Z0

N255M01

(T06 .242 DIAMETER "C" DRILL)

N260T06M06

N265G90G80G40G49

N270G54G00X.28Y8.0S996M03

N275G43Z1.0H06M08

Part 4   Machining Center Example Programs

N280G81G98Z-.350R.1F5.0

N285X19.5Y1.2

N290G80Z1.0M09

N295G91G28Z0

N300M01

(T07 .250 DIAMETER REAMER)

N305T07M06

N310G90G80G40G49

N315G54G00X.28Y8.0S458M03

N320G43Z1.0H07M08

N325G85G98Z-.29R.1F2.7

N330X19.5Y1.2

N335G80Z1.0M09

N340G91G28Z0

N345M01

(T08 7/16 DRILL TO OPEN FOR SLOTS)

N350T08M06

N355G90G80G40G49

N360G54G00X1.1Y8.5S524M03

N365G43Z1.0H08M08

N370G81G98Z-.40R.1F3.1

N375X18.0

N380Y.5

N385X10.5Y1.775

N390G80Z1.0M09

N395G91G28Z0

N400M01

(T09 7/16 DIAMETER ROUGHING END-MILL)

(RADIUS COMPENSATION D51 = .240)

N405T09M06

N410G90G80G40G49

Part 4   Machining Center Example Programs

N415G54G00X1.1Y8.5S524M03

N420G43Z1.0H09M08

N425Z.2

N430P0003M98

N435X18.0Y.5

N440P0003M98

N445X18.0Y18.5

N450P0003M98

N455X10.5Y1.775

N460G01Z-.260F10.0

N465G41X11.0D51F2.0

N470Y1.85

N475G03X10.7Y2.15I-.3

N480G01X9.55

N485G03X9.25Y1.85J-.3

N490G01Y1.7

N495G03X9.55Y1.4I.3

N500G01X10.7

N505G03X11.0Y1.7J.3

N510G01Y1.775

N515G40X10.5

N520G00Z.2

N525X20.25Y6.5

N530G01Z-.26F10.0

N535G41Y6.875D51F2.0

N540X18.25

N545G03Y6.125J-.375

N550G01X20.25

N555G40Y6.5

N560G00Z.2

N565X2.5Y-.25

Part 4 Machining Center Example Programs

N570G01Z-.26F10.0

N575G41X3.0D51F2.0

N580Y.4

N585G03X2.75Y.65I-.25

N590G01X2.25

N595G03X2.0Y.4J-.25

N600G01Y-.25

N605G40X2.5

N610G00Z1.M09

N615G91G28Z0

N620M01

(T10 7/16 DIAMETER FINISHING END-MILL)

(RADIUS COMPENSATION D52 = .2187)

N625T10M06

N630G90G80G40G49

N635G54G00X1.1Y8.5T12S524M03

N640G43Z1.0H10M08

N645Z.2

N650P0004M98

N655X18.0Y8.5

N660P0004M98

N665X18.0Y.5

N670P0004M98

N675X10.5Y1.775

N680G01Z-.26F10.0

N685G41X11.0D52F2.0

N690Y1.85

N695G03X10.7Y2.15I-.3

N700G01X9.55

N705G03X9.25Y1.85J-.3

N710G01Y1.7

Part 4 Machining Center Example Programs

N715G03X9.55Y1.4I.3

N720G01X10.7

N725G03X11.0Y1.7J.3

N730G01Y1.775

N735G40X10.5

N740G00Z.2

N745X20.25Y6.5

N750G01Z-.26F10.0

N755G41Y6.875D52F2.0

N760X18.25

N765G03Y6.125J-.375

N770G01X20.25

N775G40Y6.5

N780G00Z.2

N785X2.5Y-.25

N790G01Z-.26F10.0

N795G41X3.0D52F2.0

N800Y.4

N805G03X2.75Y.65I-.25

N810G01X2.25

N815G03X2.0Y.4

N820G01Y-.25

N825G40X2.5

N830G00Z1.M09

N835G91G28Z0

N840M01

The following mathematical calculations are necessary to perform machining of the chamfer.

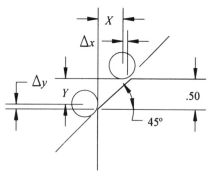

Figure 98

$$X = .5 - \Delta X = .5 - r \times \tan\frac{\alpha}{2} = .5 - .2071 = .2928$$

$$Y = .5 - \Delta Y = .5 - r \times \tan\left(45 - \frac{\alpha}{2}\right) = .5 - .2071 = .2928$$

Part 4   Machining Center Example Programs

(T11 1.0 DIAMETER END-MILL)

N845T11M06

N850G90G80G40G49

N855G54G00X-.5Y07072S229M03

N860G43Z1.0H11M08

N865G01Z-.26F10.0

N870X.2928Y10.5F1.3

N875G00Z1.0M09

N880G91G28Z0

N885G28X0Y0M05

N890M30

Subprograms for complex example program O0023

O0003

(SUBPROGRAM OF PROGRAM O0023 FOR ROUGHING SLOTS)

N1G00Z.2

N2G91G01Z-.460F10.0
(D51 = .4375/2 + .01)

N3G41Y.25D51F1.5

N4G03Y-.5J-.25

N5G01X.5

N6G03Y.5   J.25

N7G01X-.5

N8G40Y-.25

N9G90G00Z.2

N10M99

O0004

(SUBPROGRAM OF PROGRAM O0023 FOR FINISHING SLOTS)

N1G00Z.2

N2G91G01Z-.460F10.0

(D52 = .4375/2)

Part 4 Machining Center Example Programs

N3G41Y.25D52F1.5

N4G03Y-.5J-.25

N5G01X.5

N6G03Y.5J.25

N7G01X-.5

N8G40Y-.25

N9G90G00Z.2

N10M99

Milling Example

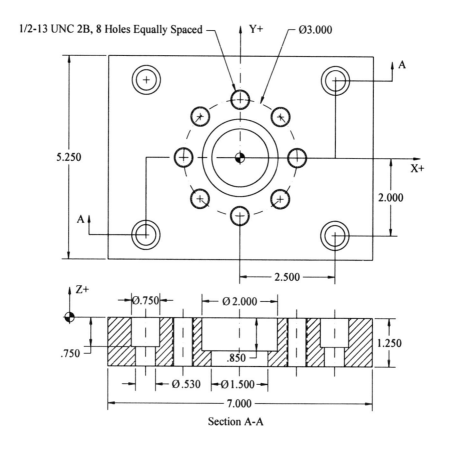

Figure 99

Part 4   Machining Center Example Programs

Example of Program No. 24

Machine: Machining Center	Program Number: O0024

Workpiece Zero

G54 X = <u>Part Center</u> Y = <u>Part Center</u> Z = <u>Top Part Surface</u>

Setup Description:

The material is steel.

Tool #	Description	Offset	Comments
T01	#6 HSS Center Drill		
T02	$\frac{17}{32}$ Diameter HSS Drill		
T03	$\frac{27}{64}$ Diameter HSS Drill		
T04	$1\frac{1}{4}$ Diameter HSS Drill		
T05	.750 Diameter 2-flute End-Mill		
T06	1.0 Diameter 4-flutes Roughing End-Mill		
T07	$\frac{1}{2}$ – 13 Tap		

Program

O0024

N10G90G80G20G40G49

N15T01M06

N20G54G00X0Y0S687M03

N25G43Z1.0H01M08

N30G81G98Z-.5F2.7R.1

N35X-2.5Y-2.0

N407P0002M98

N45X0Y-1.5

N50P0003M98

N55G80Z1.0M09

N60G91G28Z0M19

N65M01

N70M06T02

N75G90G80G40G49

N80G54G00X-2.5Y-2.0S648M03

Part 4   Machining Center Example Programs

N85G43Z1.0H02M08

N90G73G98Z-1.45R.1F3.8

N95P0002M98

N100G80Z5.0M09

N105G91G28Z0M19

N110M01

N115M06T03

N120G90G80G40G49

N125G54G00X0Y-1.5S816M03

N130G43Z1.0H03M08

N135G73G98Z-1.43R.1F4.9

N140P0003M98

N145G80Z1.0M09

N150G90G28Z0M19;

N155M01

N160T04M06

N165G90G80G40G49

N170G54G00X0Y0S275M03

N175G43Z1.0H04M08

N180G73G98Z-1.65F1.65R.1

N185G80Z1.0M09

N190G91G28Z0M19

N195M01

N200T05M06

N205G90G80G40G49

N210G54G00X-2.5Y-2.0S458M03

N215G43Z1.0H05M08

N220G82G98Z-.75R.1F3.6

N225P0002M98

N230G80Z1.0M09

N235G91G28Z0M19

Part 4 Machining Center Example Programs

N240M01

N245T06M06

N250G90G80G40G49

N255G54G00X0Y0S343M03

N260G43Z1.0H06M08

N265G01Z-1.25F50.0

N270Y.22F2.7

N275G02J-2.7

N280G01Y0F50.0

N285Z-.83

N290Y.470F2.7

N295G02J-.470

N300G01Y0F50.0

N305G00Z1.0M09

N310G91G28Z0M19

N315M01

N320T07M06

N325G90G80G40G49

N330G54G00X0Y0S343M03

N335G43Z1.0H07M08

N340G01Z-.85F50.0

N345Y.5F2.7

N350G03J-.5

N355G01Y0

N360Z-1.250

N365Y.250

N370G03J-.250

N375G01Y0

N380G00Z5.0M09

N385G91G28Z0M19

N390M01

Part 4   Machining Center Example Programs

N395T08M06

N400G90G80G40G49

N405G54G00X0Y-1.5S114M03

N415G43Z1.0H08M08

N420G84G98Z-1.3F8.7R.2

N425P0003M98

N430G80Z1.0M09

N435G91G28Z0M19

N440G28X0Y0

N445M30

Subprograms for example program O0024

O0002

(SUBPROGRAM FOR COUNTERBORES)

N1X-2.5Y-2.0

N2Y2.0

N3X2.5

N4Y-2.0

N5M99

O0003

(SUBPROGRAM FOR BOLT CIRCLE)

N1X0Y-1.5

N2X-1.0607Y-1.0607

N3X-1.5Y0

N4X-1.0607Y1.0607

N5X0Y1.5

N6X1.0607Y1.0607

N7X1.5Y0

N8X1.0607Y-1.0607

N9M99

Part 4 Machining Center Example Programs

Example Illustrating Application Of Function With Radius Compensation

Machine: Machining Center	Program Number: O0025

Workpiece Zero

G54 X = Lower Left Corner Y = Lower Left Corner Z = Top Part Surface

Setup Description: The material used is free cutting brass .100 thick.

SFM = 300

S = 6100 RPM

Feed per one flute = .0015

$F = 4 \times .0015 \times 6100 = 36$

Tool #	Description	Offset	Comments
T01	.1875 Diameter 4-flute HSS End-mill		

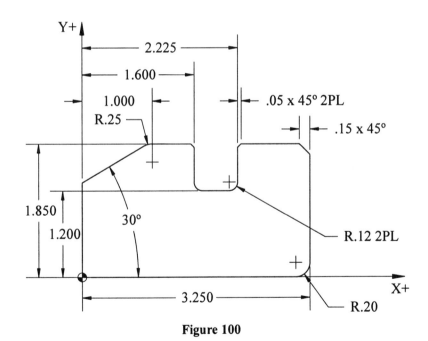

Figure 100

Note: The example given here is limited to a finishing pass only. Work holding has not been taken into consideration. In order to write a program, follow the necessary mathematical calculations shown below:

Part 4 Machining Center Example Programs

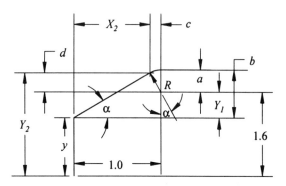

Figure 101

$$Y = 1.6 - Y_1 \quad Y_1 = b - a$$

$$\tan \alpha = \frac{b}{1.0} \quad b = 1.0 \times \tan \alpha = 1.0 \times \tan 30° = .5773$$

$$\cos \alpha = \frac{R}{a} \quad a = \frac{R}{\cos \alpha} = \frac{.25}{\cos 30°} = .2886$$

$$Y_1 = b - a = .5773 - .2887 = .2887$$

$$Y = 1.6 - Y_1 = 1.6 - .2887 = \underline{1.3113}$$

$$Y_2 = 1.6 + d$$

$$\cos \alpha = \frac{d}{R} \quad d = R \times \cos \alpha = .2165$$

$$Y_2 = 1.6 + d = 1.6 + .2165 = \underline{1.8165}$$

$$X_2 = 1.0 - c$$

$$\sin \alpha = \frac{c}{R} \quad c = R \times \sin \alpha = .125$$

$$X_2 = 1.0 - c = 1.0 - .125 = \underline{.875}$$

Example of Program No. 25

Program

O0025

N10G90G80G20G40G49

N15G54G00X-.2Y-.2S6100M03

N20G43Z1.0H01M08

N25G00Z-.1

N30G01G41X0D31F36.0

N35Y1.3113

N40X.875Y1.8165

N45G02X1.0Y1.85I.185J-.2165

N50G01X1.55

Part 4 Machining Center Example Programs

N55X1.6Y1.8

N60Y1.3

N65G03X1.7Y1.2I.0

N70G01X2.125

N75G03X2.225Y1.3J.1

N80G01Y1.8

N85X2.275Y1.85

N90X3.1

N95X3.25Y1.7

N100Y.2

N105G02X3.05Y0I-.2

N110G01X-.02

N115G40Y-.2

N120G00Z1.M09

N125G91G28Z0M05

N130G28X0Y0

N135M30

Example Program For Drilling Of 1000 Holes Using Only Six Blocks Of Information

Machine: Machining Center	Program Number: O0026

Workpiece Zero

G54 X = <u>Lower Left Corner</u> Y = <u>Lower Left Corner</u> Z = <u>Top Part Surface</u>

Setup Description: The material used is 4140 steel.

$$S = \frac{12 \times V}{\pi \times D} = 3307$$
$$F = .002 \times 3307 = 6.614$$

Tool #	Description	Offset	Comments
T01	.052 Diameter HSS Drill		SFM = 45

Part 4 Machining Center Example Programs

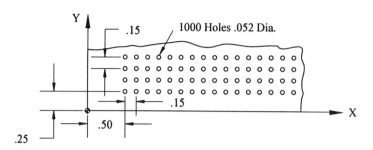

Figure 102

In the illustration above the grid of holes shown represent only a portion of those on the part. The actual part contains 4 rows of 250 holes each.

Example of Program No. 26

    O0026

    N001G90G00G80G40G49G54X.5Y.25S3307M03

    N002G43P0008L4Z1.0M98H01

    N003G91G28X0Y0Z0M30

    O0008

    N001G91G81G98X.15Z-.05L250R.05F6.6M08

    N002G80G00X-36.15Y.15

    N003M99

Example Illustrating Application Of Mathematical Formulas
The cutter used is 1/2 diameter.

Example of Program No. 28

    O0028

    N10G90G80G20G40G49

    N15G54G00X-.5Y-.5S600M03

    N20G43Z1.0H01M08

    N25G01Z-.5F5.0

    N30X-.25F3.6

Part 4 Machining Center Example Programs

N35Y5.25

N40X0

N45G03X.25Y5.5J.25F2.0

N50G01Y5.9036F3.6

N55X.5964Y6.25F2.5

N60X2.8554F3.6

N65X3.8891Y5.768F2.5

N70G02X4.25Y5.2015I-.2641J-.5665F2.0

N75Y1.8249F3.6

N80X3.5Y1.552

N85Y.4203F3.6

N90X3.3541Y-.1241F2.5

N95X2.8844Y-.25

N100X-.25F3.6

N105G00Z1.M09

N110G28Z1.0

N115G28X-.2500Y-.2500M05

N120M30

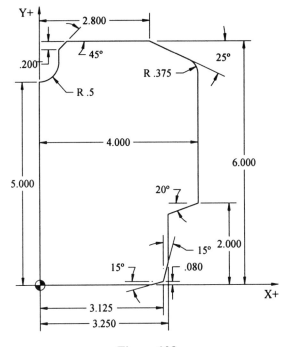

Figure 103

Notes: Automatic return to zero return is performed in the absolute system. The following is an explanation of the more difficult parts of the program.

N50G01Y5.9036F3.6

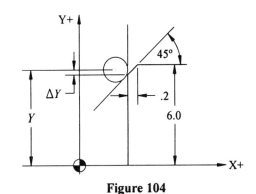

$Y = 6.0 - .2 + \Delta Y$

$r = .25 \quad \alpha = 45°$

$\Delta Y = r \times \tan\left(45 - \dfrac{\alpha}{2}\right) = .1036$

$Y = 6.0 - .2 + .1036 = 5.9036$

Figure 104

Part 4 Machining Center Example Programs

N55X.5964Y6.25F2.5

$X = .5 + .2 - \Delta X$
$r = .25 \quad \alpha = 45°$
$\Delta X = r \times \tan \frac{\alpha}{2} = .1036$
$X = .5 + .2 - .1036 = .5964$

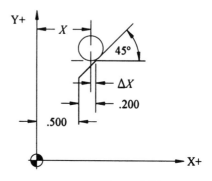

Figure 105

N60X2.8554F3.6

$X = 2.8 + \Delta X$
$r = .25 \quad \alpha = 25°$
$\Delta X = r \times \tan \frac{\alpha}{2} = .0554$
$X = 2.8 + .0554 = 2.8554$

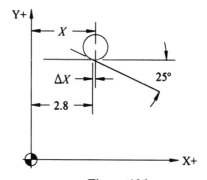

Figure 106

N65X3.8891Y5.7680F2.5

$X = 4.0 - a + \Delta X$
$R = .375 \quad \alpha = 25°$
$a = R \times \tan\left(45° - \frac{\alpha}{2}\right) \times \cos \alpha$
$a = .2165 \quad r = .250$
$\Delta X = r \times \sin \alpha = .1056$
$X = 4.0 - .2165 + .1056 = \underline{3.8891}$
$Y = 6.0 - d + b + \Delta Y$
$c = 4 - 2.8 = 1.2$
$\tan \alpha = \frac{d}{c} \quad d = c \times \tan \alpha = .5595$
$b = R \times \tan\left(45° - \frac{\alpha}{2}\right) \times \sin \alpha = .1009$
$\Delta Y = r \times \cos \alpha = .2266$
$Y = 6.0 - .5595 + .1009 + .2266 = \underline{5.7680}$

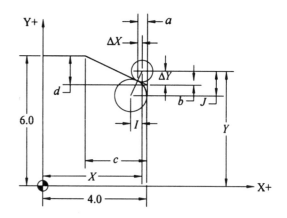

Figure 107

Part 4 Machining Center Example Programs

N70G02X4.25Y5.2015I-.2641J-.5665F2.0

$Y6.0 - d - I \quad d = .5595$

$I = R \times \tan\left(45 - \dfrac{\alpha}{2}\right) = .2390$

$Y = 6.0 - .5595 - .2390 = 5.2015$

$I = (R + r) \times \sin \alpha = .2641$

$J = (R + r) \times \cos \alpha = .5665$

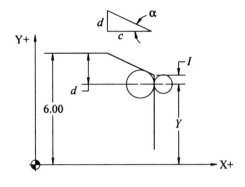

Figure 108

N75G01Y1.8249F3.6

$Y = 2.0 - \Delta Y$

$r = .25 \quad \alpha = 20°$

$\Delta Y = r \times \tan\left(45° - \dfrac{\alpha}{2}\right) = .1751$

$Y = 2.0 - .1751 = 1.8249$

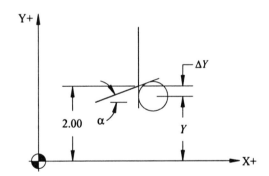

Figure 109

N80X3.5Y1.5520

$Y = 2.0 - g - \Delta Y$

$f = 4 - 3.25 = .750 \quad \alpha = 20°$

$\tan \alpha = \dfrac{g}{f} \quad g = f \times \tan \alpha = .2729$

$\Delta Y = r \times \tan\left(45° - \dfrac{\alpha}{2}\right) = .1751$

$Y = 2.0 - .2729 - .1751 = 1.5520$

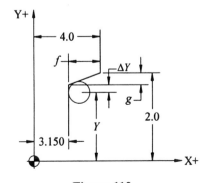

Figure 110

Part 4 Machining Center Example Programs

N85Y.4203F3.6

$Y = .08 + h - \Delta Y$

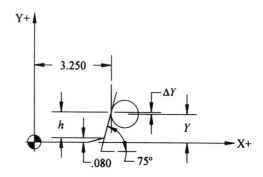

Figure 111

$\alpha = 75° \quad \tan \alpha = \dfrac{h}{K}$

$h = K \times \tan \alpha = .3732$

$\Delta Y = r \times \tan\left(45° - \dfrac{\alpha}{2}\right) = .0329$

$Y = .08 + h - \Delta Y = .4203$

Figure 112

N90X3.3541Y-.1241F2.5

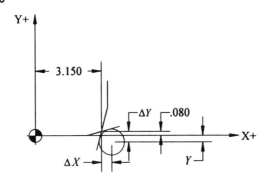

Figure 113

$Y = .08 - \Delta Y \quad \alpha = 75°$

$Y = r \times \dfrac{\cos[(\alpha + \beta)/2]}{\cos[(\alpha - \beta)/2]} \quad \beta = 15° \quad r = .250$

$\Delta Y = .2041 \quad Y = .08 - .2041 = -.1241$

Part 4 Machining Center Example Programs

N95X2.8844Y-.25

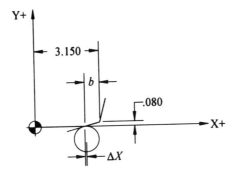

Figure 114

$$X = 3.15 - b + \Delta X$$

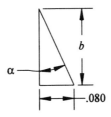

Figure 115

$$\tan \alpha = \frac{.08}{b} \quad \alpha = 15°$$

$$b = \frac{.08}{\tan \alpha} = .2985 \quad r = .250$$

$$\Delta X = r \times \tan \frac{\alpha}{2} = .0329$$

$$X = 3.15 - .2985 + .0329 = 2.8844$$

Part 4 CNC Machining Center Study Questions

Part IV
Study Questions

1. What other program word is necessary when programming G01 linear interpolation?

    a. S

    b. F

    c. T

    d. H

2. Which code activates the tool length offset?

    a. G54

    b. G40

    c. G41

    d. G43

3. When programming an arc, which letters identify the arc center location?

    a. X, Y & Z

    b. A, B & C

    c. I, J & K

    d. Q & P

4. Which of the following are modal commands?

    a. G01

    b. G00

    c. F

    d. All of the above

5. Cutter diameter compensation G41 and G42 offset the cutter to the left or the right, which command is used for climb milling?

    a. G40   c. G42

    b. G41   d. G43

6. When programming an arc an additional method exists that does not use I, J and K. Which program word is used?

    a. A   c. C

    b. B   d. R

Part 4   CNC Machining Center Study Questions

7. A block of codes at the beginning of the program are used to cancel modal commands and are called the "safety block". They are:
   a. G90G54G00
   b. G20G90G00
   c. G40G80G49
   d. G91G28G00

8. When programming an arc using the R address, a negative sign must be used with the radius value in order to create a full circle.
   T or F

9. When using the canned drilling cycle G83, which letter identifies the peck amount?
   a. Z
   b. P
   c. K
   d. Q

10. When a subprogram is used, which miscellaneous code is used to call it?
    a. M06
    b. M99
    c. M98
    d. M19

11. What character is used in the program to instate an optional block skip?
    a. (
    b.)
    c. ;
    d. /

12. What character is used in the program at the End of a Block?
    a. ;
    b. /
    c. (
    d.)

13. The letter address, O is used to identify the program number and has no other use in programming.
    T or F

14. Sequence or line numbers are identified by the letter address N. The program will not execute if they are omitted.
    T or F

15. The letter address H is used to indicate a tool length offset register number. Which preparatory function is it used in conjunction with?
    a. G54
    b. G43
    c. G42
    d. G41

315

Part 4   CNC Machining Center Study Questions

16. When using canned drilling cycles, which of the following codes are used to return the drill to the initial plane?

   a. G99   c. G90

   b. G98   d. G92

17. When using canned drilling cycles, which of the following codes are used to return the drill to the reference plane?

   a. G99   c. G90

   b. G98   d. G92

18. When using canned drilling cycles, what other letter address is necessary to identify the reference plane position?

   a. P   c. R

   b. Q   d. S

19. When using cutter diameter compensation in a program, what letter address is used to identify the location of the value of the offset?

   a. A   c. H

   b. R   d. D

20. What is the G-Code used to cancel cutter diameter compensation?

   a. G40   b. G41   c. G42   d. G43

21. Match the following definitions with the proper M-Code:

a. Program Stop   M30 ___

b. Optional Stop   M06 ___

c. End of Program   M03 ___

d. Spindle on Clockwise   M00 ___

e. Spindle on Counterclockwise   M01 ___

f. Spindle Off   M19 ___

g. Flood Coolant On   M05 ___

h. Flood Coolant Off   M98 ___

i. Spindle Orientation   M08 ___

j. Subprogram Call   M04 ___

k. Subprogram End   M09 ___

l. Tool Change   M99 ___

Part 4   CNC Machining Center Study Questions

22. Match the following definitions with the proper G-Code:

a. Linear Interpolation	G90 ___
b. Circular Interpolation Clockwise	G00 ___
c. Rapid Traverse	G81 ___
d. Dwell	G40 ___
e. Absolute Programming	G28 ___
f. Incremental Programming	G41 ___
g. Canned Cycle Cancellation	G91 ___
h. Peck Drilling Cycle	G42 ___
i. Drilling Cycle	G43 ___
j. Cutter Diameter Compensation Left	G01 ___
k. Cutter Diameter Compensation Right	G83 ___
l. Cutter Diameter Compensation Cancellation	G84 ___
m. Zero Return Command	G91 ___
n. Inch Programming	G54 ___
o. Metric Programming	G92 ___
p. Tapping Cycle	G21 ___
q. Absolute Program Zero Setting	G02 ___
r. Fixture Offset Command	G80 ___
s. Positive Tool Length Compensation	G20 ___

23. Match the following definitions with the proper letter address:

Program Number	A ___
Sequence Number	B ___
Preparatory Function	C ___
Miscellaneous Function	D ___
X Coordinate	F ___
Y Coordinate	G ___
Z Coordinate	H ___
Reference Plane Designation	I ___
Feed rate	J ___
Spindle Function	K ___
Subprogram Repeats	L ___
Tool Function	M ___
Tool Length Compensation Register	N ___
Diameter Compensation Register	O ___
Rotational Axis about the X	P ___

Part 4 Programming Numerically CMC

Rotational Axis about the Y   Q ___
Rotational Axis about the Z   R ___
Dwell in Seconds   S ___
Peck Amount in Canned Drilling   T ___
Arc Center X Axis   X ___
Arc Center Y Axis   Y ___
Arc Center Z Axis   Z ___

24. If a linear move is programmed, G01X1.5Y1.5, what is the angle of the resulting cut?

    a. 30°

    b. 180°

    c. 45°

    d. 90°

25. If a rapid traverse positioning move is programmed along the X and Y axes and the distances are unequal, the shortest distance will be achieved first.

    T or F

26. When using fixture offset programming G54-G59, it is possible to have multiple offsets in one program.

    T or F

PART 5

Computer Aided Design and Computer Aided Manufacturing (CAD/CAM)

What Is CAD/CAM?

Computer Aided Design and Computer Aided Manufacturing (CAD/CAM) utilizes computers to design drawings of part feature boundaries in order to develop cutting tool path and CNC machine code (a part program). By using CAM, the tools are defined and how they are to be used for cutting, and the specific data related to them entered. Drawing in CAD is like constructing a drawing using lines, arcs, circles and points and positioning them relative to each other on the screen. One of the major benefits of CAD/CAM is the time saved. It is much more efficient than writing CNC code line-by-line.

CAD/CAM is now the conventional method of creating mechanical drawings and Computer Numerical Control (CNC) programs for machine tools. CAD is the standard throughout the world for generating engineering drawings. The personal computer has become a powerful tool used by manufacturing for these and many other purposes. Engineers seldom use the drafting board to design their projects they now use computers extensively. Designers can create the drawings needed and share them electronically with the manufacturing department. Drawings are converted to a common file format, such as the Initial Graphics Exchange Specification (IGES) or Drawing Exchange Format (DXF). Then, the manufacturing engineer can create tool path and assign cutting tool information relative to the desired results. CAD is limited, in nature, to the generation of engineering drawings, while CAD/CAM combines both design and manufacturing capabilities. When using CAD/CAM, the drawing may be created from scratch or imported from a CAD program. It is not necessary to have the drawing dimensioned for this operation, but the full scale of the part is required. The CAD/CAM operator assigns the tools and their order of usage while creating the tool path. There are many CAD/CAM programs on the market today. The most popular ones are easy to use, have a solid background and reliability. To make good use of this computing power, it is important to fully understand the machining processes to be carried out. Just as CNC doesn't change the actual machining, the same is true of CAD/CAM for programming. Remember, the overall objective of CAD/CAM is to generate tool path for a CNC machine in the form of a CNC program. It is imperative to have a full understanding of the rectangular and polar coordinate systems. It is also necessary to have a complete understanding of cutting tool selection, speeds and feeds. Nearly all CAD/CAM programs will automatically develop speeds and feeds data based on the tool selection, however, adjustments are frequently necessary.

When constructing the part geometry, consider the type of machining operation. For instance, if the desired result is to drill a hole using a standard drill, construction of only the point that represents the hole center location in the coordinate system is necessary.

In this book, Mastercam will be featured as a CAD/CAM software example. This chapter is only intended as an introduction to CAD/CAM. Many other CAD/CAM programs use similar techniques to accomplish the same result.

Part 5   CAD/CAM

The following is a short description of the process of using Mastercam to create geometry, tool path and program code for CNC machines.

Personal Computer

The computer needed to run this type of software has important minimum requirements. Normally, a large screen is desirable for ease of viewing the geometry created. CAD/CAM programs require a lot of hard disk space so a large hard drive is also recommended. Because CAD/CAM is used to create complex drawings and perform graphic simulations, the computer has the following basic needs: The memory the computer uses to access files while working on them is called RAM, (Random Access Memory). For CAD/CAM a large amount of RAM is highly recommended (individual software manufacturers have recommended minimums). The computer's processing speed is listed in MHZ (Megahertz), again the higher the number, the better. The computer's graphics card and monitor controls the screen resolution.

Windows

To be successful using CAD/CAM, it is necessary to understand the use of a Personal Computer programs. Microsoft Windows is the most widely used operating system on personal computers. With CAD/CAM, the operator must understand the operating system and have basic skills for mouse usage, including: double-click, right mouse button and the mouse pointer. Just like most computer programs in use today, Mastercam uses a Graphical User Interface (GUI) for ease of input. The Mastercam icon menu at the top of the main menu display requires a single click with the left mouse button to activate a command. The same technique is used to activate a command from the Mastercam main menu list.

Program Startup

From the Windows main screen, look at the desktop to see if there is a shortcut icon for Mastercam Mill 8 and double-click the left mouse button. If there is not a shortcut icon, press the start button in the lower left corner with the left mouse button. Slide the mouse pointer up to Programs, to the right you

Figure 1

will see a list of all available programs. Slide the mouse pointer to find Mastercam 8 and another list will appear to the right. Again, slide the mouse pointer through the list to find the desired program. In this example, Mill 8 is used. Single click the left mouse button, the Mastercam main menu screen will be displayed as shown in Figure 2.

Many of the program functions are accessible by clicking the left mouse button while over the icon. A list of the same functions is available on the upper left of the screen in the form of a menu. The functions can be activated by using the mouse or by using short-cut keystrokes, generally the first letter of the word or, if not, they will be identified by an underscore of the letter needed. In the description that follows, the Menu selections as opposed to the Icons will be used, and the Menu item, will be **highlighted** and the short-cut keystroke underscored to match the Mastercam screen.

Part 5   CAD/CAM

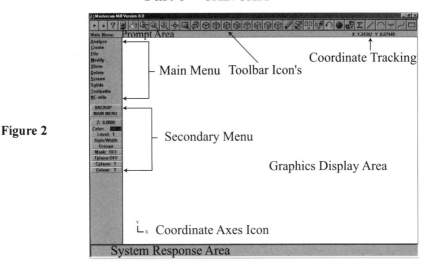

Figure 2

The Secondary Menu appears below the menu list. The top two buttons are Backup and Main Menu. Use the Main Menu button to return to main menu from anywhere within the program. When Backup is pressed, the software steps back one level in the menu. Note: The escape (Esc) key accomplishes the same as pressing Backup.

The main screen work area is black by default and can be changed to another color if necessary. Sometimes, the drawing exceeds window size. Use the Fit icon (ninth from the left on the icon bar) to fit to screen. Zoom into specific areas to magnify hard to see spots. Use the Create selection to build the part using lines, arcs, points and other commands. Use the Modify menu to clean up existing geometry.

Geometry Creation

For this example, all of the geometry shown in Figure 3 must be recreated. The first consideration when recreating geometry is where to set the origin. A zero location has

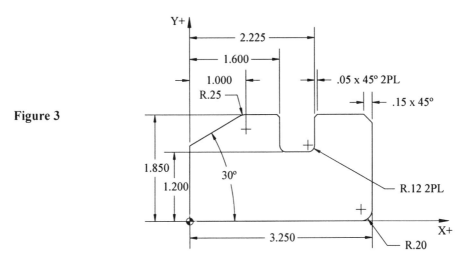

Figure 3

323

Part 5 CAD/CAM

been indicated on the print by the symbol in the lower left hand corner. It makes good sense to use this same location to start drawing.

Job Setup

Job Setup, as the name implies, is the first step that is completed in the process before the actual geometry is created. The information entered here establishes the data needed for the program such as stock size tools and part origin.

- From the Main Menu, single click the left mouse button on Toolpaths.
- To access the Job setup screen, single click on the Job setup menu item.

Explanations are given below for each item in the following illustration.

Import... allows the import of job setup information from other saved files. The Imported operations will not include part geometry because specific parameters must be set for the new geometry.

The Views... box displays all of the saved views for the active file and their origin, whether they are associative and the work offset number.

Tools... this button displays the Tools Manager list for the current file.

If no job setup or operations were imported from another file then, there will not be any tools present. Any tools that are necessary will need to be created at this time. Press the Tools button and the Tools Manager will be displayed.

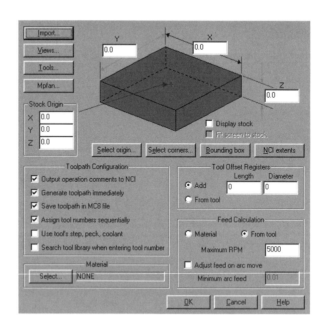

Figure 4

To add tools to the list simply right click and then select, Get from library. The Tools Manager, dialogue box will appear as Figure 5.

There is a button titled Filter... and a check box to activate the filter. If this checkbox is not checked, then all 315 tools in the standard database will be displayed in the list. By checking this box to activate this filter the number and type of tools displayed

Part 5 CAD/CAM

Figure 5

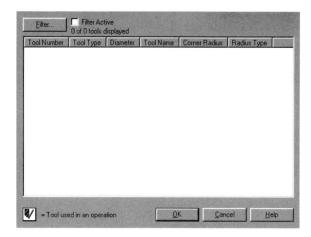

can be refined.

- Position the mouse pointer in the white area then right click.
- Choose, Get from Library.
- Use the scroll bar located on the right of the window and scroll to find a 3/16-inch flat end mill.

The chosen end mill will now be in the tool display area window of the Tools Manager as shown in Figure 6. This is the only tool needed for the program in our example. If additional tools are needed the same process may be followed to add them.

- Use the mouse to left-click to highlight, then, press OK or double-click to accept the selection. (Figure 7 will be displayed)

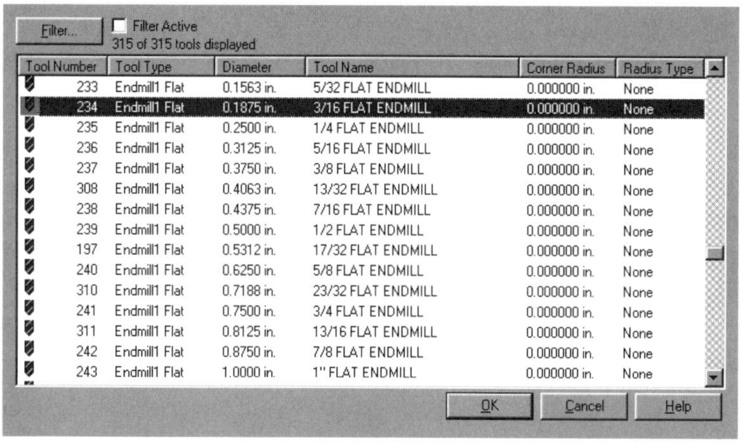

Figure 6

Stock Origin, sets the location of the stock origin. The origin for the part may be moved to a desired corner location by clicking and dragging the red arrow. The exact numerical locations may also be input into the X Y and Z to accomplish this.

X and Y Coordinates, define the outer boundary of the raw stock of the part.

Select origin, when pressed, will open the drawing file and allow the manual selection of the stock origin with the mouse directly from the drawing.

Part 5   CAD/CAM

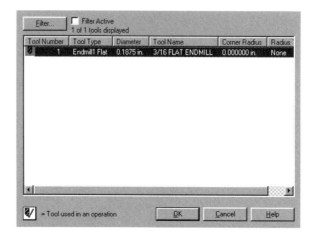

Figure 7

Select corners..., when pressed, also opens the drawing file and allows the manual selection of the stock corners directly from the drawing.

Bounding box, defines the actual limits of the stock size by finding the extents of the selected geometry.

NCI extents, will automatically set the stock boundaries by calculating the necessary size dependant upon the tool movements in the NCI file.

Even though no drawing exists yet we can establish the origin and stock size now.

- Use the left mouse button and click on the arrow in the center of the part, then click the mouse button over the lower left hand corner.

The arrow indicating the location of the part origin will move to the corner selected.

- Now key in -.100 for both the X and Y stock origin.

This will allow for the .100 thousandths excess material on all contours.

- To set the stock size, key in the dimensions for the X, Y and Z including raw material (the Z value must be positive).

Toolpath Configuration, allows for the items shown when checked. Material can be selected from a list when the select button is pressed. This selection has an effect on the feed and speed calculations. Display Stock, when checked, shows the raw material boundaries within the drawing file as defined in Job setup.

- Check the Display stock check box.

Fit screen to stock, when checked, allows the maximum viewable area to be used to display the stock.

Tool Offset Registers, are used to determine the tools length and diameter offsets.

Feed Calculation, is allowed based on the tool material, or the work material and a maximum RPM may be set.

Mastercam Feature Creation Step-by-Step

Please note, the steps given here are by no means the only method by which this geometry may be created. Individual preferences and speed will ultimately be the

Part 5   CAD/CAM

determining factor in how drawings are created.

To create the drawing, follow these steps:

- From the Main Menu, position the mouse pointer over **Create** and press the left button, or press the letter "C" to activate the create menu.

Note: Even though many times the short-cut keystroke letter is in uppercase it is unnecessary to key it that way. In other words, a lower case "c" will accomplish the same thing.

- By the same method, choose **Line**, or press the letter "L" to activate the line menu
- Choose **Horizontal**, or the letter "H" to activate the horizontal line menu

In the system response area the prompt reads: "Create line, horizontal: Specify the first endpoint" (The system defaults to **Sketch**, under Point Entry).

- Key in the coordinates for the start point of the line: 0, 0 and press Enter (where 0, 0 = X value, Y value) (the Z value is zero unless otherwise specified on the secondary menu)

The first point is drawn and a horizontal line is attached to it. As the mouse is moved the line stretches like a rubber band to the point. In the system response area the prompt reads: Length = (value of the line length). The default mode is Sketch so whenever the mouse is moved the line length changes correspondingly.

Note: If the coordinate values are known, begin typing the values, and the software automatically enters that mode and the insertion point window will pop-up where the values are keyed in.

- Key in: 3.250, 0 for the second end point and press Enter
- Press Enter again to accept the Y coordinate of zero

Use the mouse to position the cursor over the Secondary Menu button labeled "Backup" (just below the Main Menu) and press it.

Pick **Vertical**

In the system response area the prompt reads: "Specify first endpoint".

- Pick **Last**, to use the ending point of the last line entered
- Key in 3.250, 1.850 Enter, Enter (the second time is to accept the X coordinate of 3.25)

The vertical line is drawn.

- Press Backup (use the Esc key to Backup one level each time pressed)
- Pick **Horizontal**
- **Last**
- Key in .625 + 1.6, 1.85, press Enter and the line will be drawn.
- Press Enter again to accept the Y coordinate of 1.85.

Note: Because the absolute coordinate in X for the last point was not given, a calculation was necessary. The software allows for mathematical calculations within the

Part 5   CAD/CAM

prompt area following the Algebraic Order System (AOS).

- Now select Backup or press Esc.
- **V**ertical
- **L**ast
- Key Y1.2 and press the enter key.

Note: If there is no need to change the X coordinate, key Y and the value, then only Y changes.

The system prompts for entry of the X coordinate of 2.225

- Press Enter to accept
- Press Backup or the Esc.
- Choose **H**orizontal
- **L**ast
- Key X1.6 Enter
- Enter to accept the Y coordinate of 1.2
- Press Backup or Esc.
- **V**ertical
- **L**ast
- Key Y 1.850, Enter
- Enter to accept the X coordinate of 1.6
- Press Backup or Esc.
- **H**orizontal
- Last
- Key X 1.0 and Enter

Note: Because the start of the arc is known to be that location in "X". Enter to accept the Y coordinate of 1.85.

- Press Backup or Esc.
- **P**olar

In the system response area the prompt reads: "Create line, polar: specify an endpoint".

- Key in 1.0, 1.6 (the coordinate value of Y is obtained by subtracting the known radius value of .250 from 1.85 = 1.6)

In the system response area the prompt reads: "Enter the angle in degrees".

- Key in 120° and press Enter (because the angle of 30° is given on the print and when the 30° is added to 90° it = 120°) see Figure 8.

In the system response area the prompt reads: "Enter the line length".

- Key in .25 Enter

Part 5   CAD/CAM

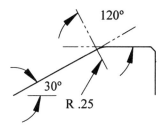

Figure 8

- Select Backup or press the Escape key twice
- Select **A**rc
- Select Endpoints

In the system response area the prompt reads: "Arc, endpoints: Enter the first endpoint".

- Pick the upper end of the line just drawn and left click the mouse to accept

In the system response area the prompt reads: Enter the second endpoint".

- Enter the second end point by positioning the mouse over the required endpoint

Note: If the Auto Cursor is ON, a box will be drawn when the cursor is positioned over the endpoint. The Auto Cursor may be turned ON or OFF by pressing the right mouse button and pressing the left mouse button to check mark it to ON or OFF. When this function is ON, it is useful for simple drawings, but may be annoying on complex drawings.

- Click the left mouse button when the endpoint desired is found

In the system response area the prompt reads: "Enter the radius".

- Key in .25, Enter

Upon doing this two full circles will be drawn.

In the system response area the prompt reads: "Select an arc".

- Carefully move the mouse pointer to select the portion of the arc that meets the print requirement and left click when ready
- Main Menu
- **C**reate
- **L**ine
- **T**angent
- **A**ngle

In the system response area the prompt reads: "Create line, tangent at an angle: select an arc or spline".

- Click the left mouse button on the arc

In the system response area the prompt reads: "Enter the angle in degrees".

Part 5 CAD/CAM

• Key in 210° and then Enter (because 180° + 30° = 210°)

In the system response area the prompt reads: "Enter the line length".

• Key in 2.0 and then Enter (because this length is sufficient to intersect with the vertical line from X Y zero) and a line 2 inches long will be drawn in both directions from the tangency point

In the system response area the prompt reads: "Select Line to Keep".

• Pick the line to the left of the arc tangent
• Main Menu
• **Create**
• **Line**
• **Vertical**
• Key in 0,0
• Key Y 1.85, Enter and Enter again to accept the X coordinate of zero

To trim the excess lines and fillet the corners as specified on the print

If the drawing exceeds the display area of the screen, press the -.8 Zoom Icon

• Main Menu
• **Modify**
• **Fillet**

In the system response area the prompt reads: "Fillet angle is less than 180°; trim to fillet, fillet radius = 0.2500 fillet: select an entity". (The default fillet size is 0.2500 radius).

• Press **Radius**
• Key in .2 and press Enter

Note: The Angle should be <180 S and trim should be set to Y (Yes)

In the system response area the prompt reads: "Fillet angle is less than 180; trim to fillet; radius .200. Fillet: Select an entity".

• Use the mouse to select an entity, Choose Line 1

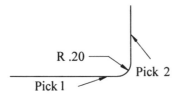

Figure 9

• Then choose line 2 as shown in Figure 9
• While still in fillet mode, change the radius to .12 for the 2 other fillets

Part 5 CAD/CAM

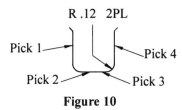

Figure 10

- Pick each vertical and horizontal line as shown in Figure 10

All fillets are now done

Now trim the excess lines off from the left angular and vertical lines.

- Main Menu
- **Modify**
- **Trim**
- **2 entities**

In the system response area the prompt reads: "Trim (2): select the entity to trim".

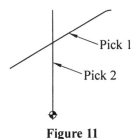

Figure 11

- Choose the angular line, then the last vertical line as shown in Figure 11

The excess lines will be trimmed.

Now the chamfers must be created.

- Main Menu
- **Create**
- **Next Menu**
- **Chamfer**
- **Distances** (the default value for distances is .25, change as needed)

Set both distances at .05.

- Key in .05 and press the Enter key and then again key in, .05 and press the Enter key for the second distance

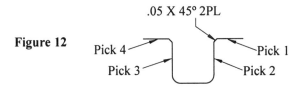

331

Part 5 CAD/CAM

- Pick the necessary lines as shown in Figure 12
- Select **Distances** again and reset each at .15

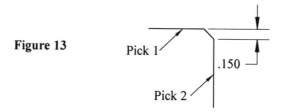

Figure 13

- Pick the necessary lines as shown in Figure 13

The construction line that was used for the creation of the .25 arc, tangent to the angular line, should be deleted now

Main Menu

- **Delete**

Use the mouse to position the cursor over the construction line and the press the left mouse button and the line will be deleted

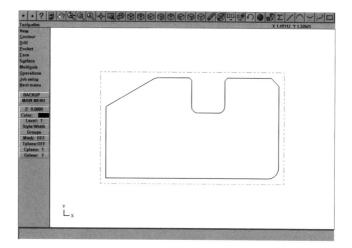

Figure 14

The drawing is complete and the file should be saved.

- Main Menu
- **File**
- **Save**

A windows dialogue box comes up titled "Specify File Name to Write".

- File Name: key in the name of the file
- Press the Save button

Tool Path

Once the geometry has been created, tool paths can be generated

- From the Main Menu select **Tool paths**

Part 5   CAD/CAM

Choose the type path that applies to the geometry. For this example, chose contour.

• Press **C**ontour

The prompt area at the top left of the screen states: Contour: select chain 1.

Note: The default setting is chain, so the contour may be picked without first activating the chaining mode.

In the system response area the prompt reads: "Chaining mode: Full Chaining mask None"

• Use the mouse to position at an acceptable starting location for the tool and left-click the mouse button

This selection determines where the actual tool path will begin, so take care to select an appropriate point. An arrow on the drawing contour will be displayed indicating the direction of the chain. The chain direction determines the tool travel direction. Tool offset direction is always dependant on chaining. For instance, if the arrow is point-

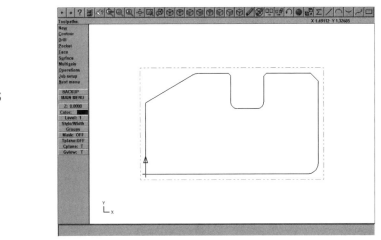

Figure 15

ing up, as shown in Figure 15, the tool will travel in the Y positive direction on the part geometry.

If the arrow direction is not correct, press **R**everse to change the direction. The offset direction from the contour determines climb or conventional cutting. Press **R**everse again and the direction toggles. If the starting location is not the desired location, press **move F**wd or **move B**ack until the desired starting position for the tool is found.

• Press **D**one

The dialogue box, "Contour (2D) – C:\MCAM8\MILL\NCI\PART V FIGURE 3.NCI - MPFAN", will pop-up as shown below in Figure 16.

Notice that the .187 diameter flat end mill that was identified earlier in the Job setup is present. If additional tools are required they can be added from this screen by following these directions: Left-click on tool to select; right click to edit or define new tool. Right click in the large white area under the statement. Upon right click, three

Part 5   CAD/CAM

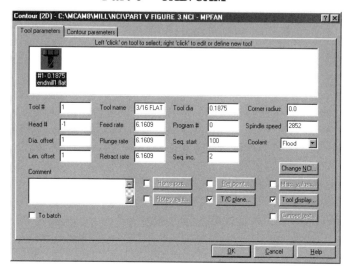

Figure 16

choices become available: Get tool from library, Create new tool, Get operations from library, or Job Setup.

Choose Get tool from library, the Tools Manager, "C:\MCAM8\MILL\TOOLS\TOOLS.TL8", dialogue box will come up. Select the needed tool from the list.

Contour (2D)

The contour (2D) window has two tabs: Tool parameters and Contour parameters. The default tab is Tool parameters.

Tool Parameters

The Tool Parameters tab has several other items that can be adjusted by the programmer.

The Tool #, defaults to number 1 but can be changed to any number desired. Head #, refers to the magazine where the tool is mounted in the case of multiple tool magazines. The Dia. offset, defaults to 41 but it also can be changed to any number, dependant upon the available pockets in the tool changer of the machine being programmed. For instance, if there are 30 pockets for tools the logical diameter offset number for tool number 1 would be 31.

Length offset, defaults to 1 on the first selected tool and sequentially after that. The length-offset number will normally correspond with the tool number but any number may be used.

Feed rate is the linear feedrate at which the tool will travel and is determined by the tool material selected. This feed can be modified in this box.

Plunge rate, is the linear Z axis feedrate at which the tool will travel determined by the tool material that is selected. This feed can be modified in this box.

334

Part 5 CAD/CAM

Retract rate, is the speed at which the tool will retract to the reference plane at completion of the cutting cycle. This feed can be modified in this box.

Tool dia defaults to the selected tools diameter.

Program #, a specific program number may be entered here.

Seq. start, is the starting number for the line sequence "N" numbers in the program. Any starting value can be entered here. Common starting numbers are 1, 2, 5, 10 or 100.

Seq. inc. is the sequence number increments for the line sequence numbers. Any value may be entered here. Common increments are 1, 2, 5 or 10.

Corner radius is the value of the corner rounding around corners on a profile. Adjusting this value can eliminate some deburring operations.

Spindle speed can be set at the RPM desired for each tool.

Coolant allows for the use of flood coolant, Mist coolant, (if available) through the Tool coolant or OFF.

The Comment section allows for comments to be inserted within the program for operator guidance.

Home pos allows for setting of a specific tool change position.

Ref. point is used to adjust Approach and Retract tool path intermediate points.

Misc. values can be used to set up to ten integer values. The most commonly adjusted integers are; Work coordinates, (G92 or G54's), Absolute/Incremental (0=ABS) and Reference return, (G28 or G30).

Rotary axis parameters are used to create toolpath motion where a rotating axis is used for cylindrical parts.

T/C plane opens the Tool/Construction Plane dialog box. By using this box the planes for construction and machining, origins, and work offset for the toolpath can be set.

Tool display parameters are used to manipulate how the toolpath appears in the graphics window.

Figure 17

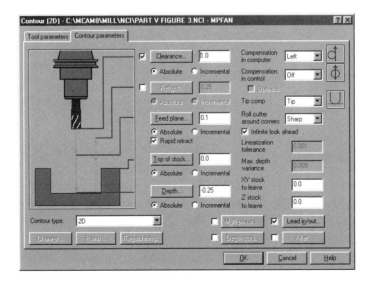

335

Part 5   CAD/CAM

Contour Parameters
• Press the Contour parameters tab.

Input the appropriate values for clearance, retract, feed, top-of-stock and depth as required. In this example, clearance is 1-inch, feed is set at .1 in absolute, top-of-stock is 0 in absolute and depth is -.25 in absolute see Figure 17.

The button labeled Depth Cuts enables multiple passes for the Z step amount. A maximum roughing step can be established, the number of finish cuts, the finish step and the amount of stock to leave. When checked, the second button, Multiple Passes, enables roughing and finishing passes in the X Y direction with control of finish passes at final depth or all depths. The lead in/out button enables entry and exit lines and arcs into the cut relative to the direction of the chain mentioned earlier.

The selection box Compensation in Computer, determines the offset direction of the cutter in the computer (be sure the correct direction is indicated, for this example, it should be left). If Compensation in Control is ON, the post processor will develop a diameter compensation number specific to the tool and insert a G41 or G42 and a D# into the code of the program.

Figure 18

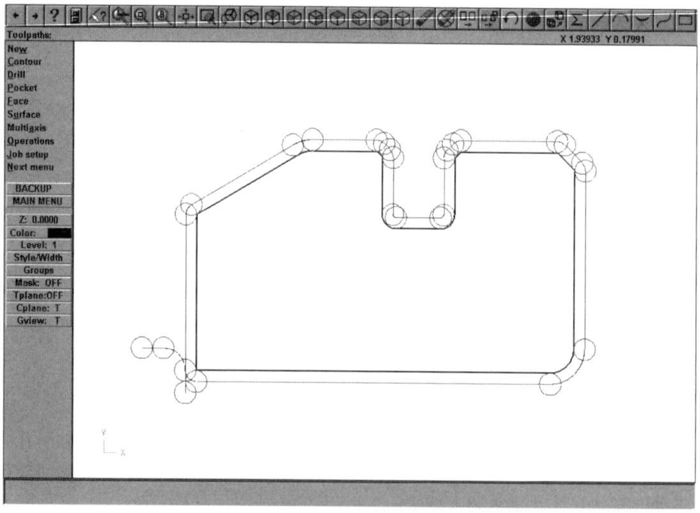

Press **O**perations

Figure 19

Part 5   CAD/CAM

• Press OK.

The tool path will be drawn on the screen as shown in Figure 18.

• Press Back Plot, and then press run, to observe a simulation of the cutters actual path. There are multiple options available here for the display of the type of tool path, the tool holder and more. It is also helpful to set the view to isometric for better visualization.

Verification

Press Verify for a solid model simulation of the part being cut.

Press the icon of the running man to simulate machining quickly, or the walking icon to simulate machining slowly.

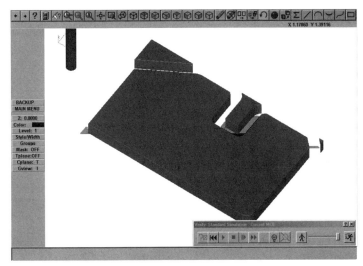

Figure 20

Press the play button, the single arrow pointing to the right, to run the simulation. At this point, if all looks well, Press the x to close the floating window.

Post Processing

The last thing to do is to post process the CNC code, for the contour. This converts the tool path just created into CNC code that the machine tool can read in order to machine the part.

• Press post.
• Run the post processor?
• Answer Yes.
• Specify the file name to write in the dialogue box, the default file name is the same as the tool path file name, if a different name is desired, change it now.

Part 5   CAD/CAM

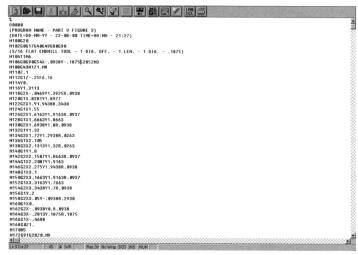

Figure 21

- Press save, and in a matter of seconds the program will be post processed and will open the program file editor as shown in Figure 21.

There are separate post processors designed for each different type controller, the default controller type is a Fanuc style G-code. Close this screen by pressing the close window button or choosing file then exit, press OK. To close the operations manager window main menu file save. Once the file is post processed, it is ready to be sent to a specific machine controller. Check the manufacturer's manual for specific directions for this procedure.

After the completion of all the items mentioned above, the programmer can print out a set-up sheet from the NC Utilities menu function. This, in effect, is a form of automated process planning and this document can be used by CNC setup persons to aid in the machine setup.

Associativity

The concept of associativity is inherent to most modern CAD/CAM programs. The information input to the program regarding the tool path, tool, material, and parameters specific to each, are linked to the geometry. This means that if any of the parameters for the parts mentioned above are changed, the other related data can be regenerated, to take these changes into account, without recreating the entire operation.

CAD/CAM is the tool of choice for creating CNC programs and the power it has now will only be magnified in the future. The basic concepts that were demonstrated here are merely a taste of the capabilities CAD/CAM has to offer.

PART 6

Conversational Programming

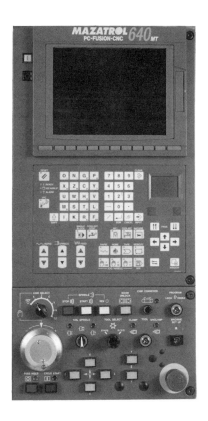

What is Conversational Programming ?

For many years, the concept of programming the CNC machine tool at the controller was thought of as inefficient and tedious. When orders of a small lot size were to be produced, the choice was almost always manual machines. Today, this is not the case, largely because of the advances in conversational programming.

Conversational programming is becoming more widely used throughout the industry and is available as standard on many machine tool controllers. Its major advantage is that it gives the machinist the ability to write programs at the machine quickly and easily. Typically, the process includes a sequence of questions the machinist/programmer must answer, sometimes called "question answer format" or "prompting". As these questions are answered, the program is constructed. Most controls with this capability also allow the machinist/programmer to graphically check the tool path to verify the program. If the program has flaws, the controller will not execute the tool path and the programmer must remedy the problem. When program errors do occur, an alarm number will show on the screen indicating what the problem is and where, in the program, it occurred. This is obviously a better method of finding errors than actually running a part. Another capability of conversational programming is its feature to perform calculations when programming data are missing from the engineering drawing. The programmer constructs intersection points or tangency points and with this information the controller computes the desired geometry. Feeds and speeds are automatically calculated based on the workpiece material and cutting tool material. The data necessary to do this is stored in the controller memory in the cutting condition parameters.

POINT MACH-ING	LINE MACH-ING	FACE MACH-ING	MANUAL PROGRAM	OTHER	WPC	OFFSET	END	SHAPE CHECK

Figure 1

Conversational "shop floor" programming uses the concept of operator prompting combined with a Graphical User Interface (GUI). Questions throughout the programming process prompt the user for information necessary to complete the part program. Icons accessed through the function buttons on the controller identify machining operations, i.e. Point Machining, Line Machining and Face Machining etc. see Figure 1.

Much like CAD/CAM, the process resembles recreation of the part geometry by constructing the shapes using lines, arcs, and points combined with other features.

This information is used in combination with Tool Identification Parameters and Cutting Condition Parameters to generate the tool path code needed to control the

Part 6   Conversational Programming

machine. The programmer has the added functions of the controller's ability to calculate for unknown coordinate values and have them automatically inserted into the program where they are needed. In most cases, no calculator is needed for trigonometric calculations. These constructions resemble the formerly popular Automatic Programmed Tool (APT) method of programming. The actual program the machine executes is still a G-Code format but the operator may never see the actual code on the display.

This type of programming combined with the ability to call G-Code sub programs has tremendous power. Mazatrol, a conversational programming system offered on all MAZAK machine tools allows G-Code similar programming within its conversational language in what is called a Manual Programming Process. The acronym used to call this type of program process for Mazatrol is MNP, where MN stand for Manual and P for Programming.

There are many different conversational languages for programming available, and one of the complaints is the lack of standardization between the various machine tool builders. The industry leader in conversational programming has always been MAZAK with their Mazatrol language. The focus of this chapter will be centered on this language only. Other languages contain similar techniques that accomplish nearly the same result.

The sequence of events followed by the programmer to create a Mazatrol program are very similar to those used in manual programming. Study of the technical part drawing, work holding considerations and tool selection must take place prior to preparation of the part program. Once the program is completed, the tool path must be verified by graphical simulation. At this point, if all checks well, the operator takes over for the measuring of tool and work offsets and one final program test by dry run. And, finally, the first part of CNC machining begins.

Turning Center Program Creation

Following, are brief descriptions of the programming process for Turning Centers.

The control must be in the program-editing mode and a work number (program number) must be identified. Note: There is no need for the letter address O to precede the program number with Mazatrol programs. Before any programming can take place, the type of program to create either Mazatrol or EIA/ISO must be determined. All MAZAK machines use Mazatrol as their standard with EIA/ISO (G-Code) as an optional feature. The Turning Center programs are made up of these four basic parts; a Common Data Process, Machining Process, Sequence Data and an End Process.

Common Data Process

The information at the head of the program applies to the entire program. The programmer is prompted to answer the following questions for this common data.

Part 6   Conversational Programming

What is the material?

The controller is preset with standard materials of Carbon Steel, Alloy Steel, Cast Iron, Aluminum and Stainless to choose from. This choice affects the automatic calculation of cutting feeds and speeds throughout the program. It is possible to add other materials to the cutting condition parameters if the material needed is not available.

What is the maximum outer diameter of the workpiece?

This value is dependant upon the diameter geometry of the raw workpiece. If the programmer inadvertently exceeds this diameter with any values in the program, the controller will set off an alarm preventing execution of tool path verification and automatic operation.

What is the minimum inner diameter of the workpiece?

If the workpiece geometry is of solid bar stock, this value may be set to zero. If an inner diameter exists, the programmer must input this value. Doing this prevents the generation of tool path where material is nonexistent.

What is the overall length of the workpiece?

The overall length of the workpiece along the Z axis, including the clamping amount should be entered for this value. If a programmed value exceeds this length, the controller will set off an alarm preventing execution of tool path verification and automatic operation.

Would you like to limit the spindle RPM to a maximum?

This enables the programmer to limit the spindle RPM to a predetermined amount. If no value is input into this data slot, the controller will execute the maximum spindle RPM when at the centerline in the X axis. This maximum RPM may be undesirable in some cases.

How much material would you like to leave for a finish pass in the "X" axis?

The amount of material to be left for a finishing pass in the X axis is input at this time. This value is input in consideration of the diameter of the workpiece. For example: if a value of .040 inch is input, the amount taken off the diameter is = to .080 inch for the finishing pass.

How much material would you like to leave for a finish pass in the "Z" axis?

The amount of material to be left for a finishing pass in the Z axis is input at this time.

Part 6   Conversational Programming

What is the amount of material taken off the work face to attain Z zero?

It is common to machine material from the face of the workpiece in order to attain a finished surface that establishes the Z zero for the part. This amount is dependant upon the condition of the material and programmer preference.

Machining Process

Here, the individual machining process data are identified in order to complete the workpiece definition. In other words, what type of machining is to be done? In the figure below, note the choices are BAR, CPY, CNR, EDG, THR, GRV, WORKPIECE SHAPE and END.

Figure 2

Bar (BAR) machining is used for outside diameter (OD) or, inside diameter (ID) turning and boring. Copy (CPY) machining is used for OD or ID machining of existing geometries like castings or forgings, where a uniform amount of material is to be removed and is other than solid bar stock. Corner (CNR) machining is used when additional cutting tools are needed to finish corners that cannot be cut because of tool geometry limitations. Edge (EDG) machining is used to perform machining on the face of the workpiece. Thread (THR) is for machining of external and internal screw threads. Grooving (GRV) is machining of external and internal grooves. Workpiece Shape machining is similar to CPY except that the material removal shape does not need to be uniform. END is used to end the program. The arrows at the right offer some additional options of Drilling, Tapping and Manual Programming (as mentioned above like G-Code within the Mazatrol program).

Once a selection is made for the type of machining operation, then more information is needed to identify how to apply it. In the figure below the Bar machining has been selected and new choices are displayed. Those items that are bold in the graphic are captured-type of cuts. The first on the left, OUT is used to perform general OD machining and the third from the left, IN is used for general ID machining like boring.

Figure 3

For Bar Machining, the related information are: Feeds and Speeds, which are automatically calculated by pressing a function key and are based on parameter information directly associated to the selected cutting tool and material; Tool Selection for roughing and finishing cycles; The Starting Point in X, The Starting Point in Z, The Finish Point in X and, the Finish Point in Z.

Part 6   Conversational Programming

Sequence Data

The finished workpiece shape is identified by input of point data until the desired geometry exists using lines (LIN), tapers (TPR), arcs, chamfers and fillets, limited only by the tool configuration. The same type data are necessary for internal bar machining. In the figure below, the two types of arcs shown represent convex and concave shapes respectively and the CENTER command is needed to identify the arc center point.

Figure 4

When all the geometric data are entered and the shape is defined properly, then the SHAPE END function key is pressed to end the process. The remainder of the program is constructed in the same manner until complete.

End

This process ends the program (similar to M30 in G-Code programming) and offers the opportunity to the programmer to: set a counter for the number of workpieces to be machined, set the return position of the turret after machining ends; identify a next program number to machine, whether to continue the same program repeatedly or not, and how many repetitions; also, the shift amount for the coordinates system can be set.

Shape Check

Once the program data are entered, the programmer may perform a Shape Check of the geometry to verify its accuracy. The controller will not allow the shape check if there are serious problems with the geometry and an alarm will be displayed on the screen identifying the program number, process number and sequence number of the mistake. Performing a shape check will draw the finished workpiece geometry on the screen in two-dimensional form.

Tool Path Verification

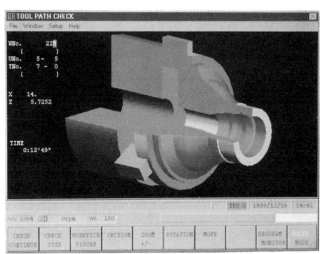

Figure 5

Part 6   Conversational Programming

Another step is Tool Path Verification and may be used to check the geometry and tool path. The newest controllers are equipped with solid model rendering of the workpiece raw stock configuration. The programmer can simulate on-screen, the actual machining of the workpiece in real-time by pressing check continue.

If an area of the graphic is hard to see because of its size, the programmer can Zoom into that area and magnify it for better viewing. Tool shapes are graphically simulated as well, allowing an excellent visual aid for correcting any problems.

Often, a workpiece has internal features that are difficult to see even with solid modeling. The newest controls have the graphical capability to section the workpiece, allowing a full visual representation that offers even more assistance to the programmer for verifying programs.

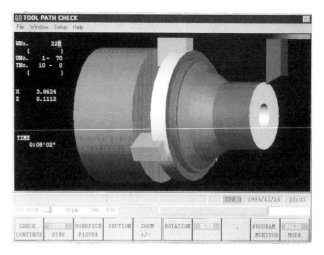

Figure 6

Once these verification steps are complete, the machinist may begin the first article of CNC automatic operation.

Machining Center Program Creation

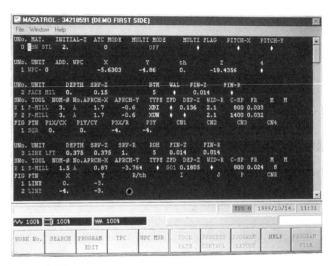

Figure 7

Part 6   Conversational Programming

Following are brief descriptions of the programming process for Machining Centers. The control must be in the program-editing mode and a work number (program number) must be identified. Note: There is no need for the letter address O to precede the program number with Mazatrol programs. Before any programming can take place, the type of program to create either Mazatrol or EIA/ISO must be determined. All MAZAK machines use Mazatrol as their standard with EIA/ISO (G-Code) as an optional feature. The construction of a Mazatrol Machining Center program contains these basic parts: a Common Data Unit, identification of a Coordinate System, the Machining Units and their Sequence Data and an End unit.

Common Data Unit

The information at the head of the program applies to the entire program. The programmer is prompted to answer the following questions for this common data.

What is the material?

The controller is preset with standard materials of Cast Iron, Ductile Cast Iron, Carbon Steel, Alloy Steel, Stainless Steel, Aluminum and Copper Alloy to choose from. This choice, combined with tool material, affects the automatic calculation of cutting feeds and speeds throughout the program. It is possible to add other materials to the cutting condition parameters, if the material needed is not available.

Where is the Initial Z position?

This value identifies where all of the tools will move to, at rapid traverse, before machining begins. In G-Code programming, this is the same as the Reference Plane (R-Plane). A common reason for setting this at a particular height is to provide for clearance of work holding clamps.

Where should tool changes take place, ATC Mode

Automatic Tool Change (ATC) mode establishes how the tool is returned to position for a tool change. For example, one selection first returns the Z axis to the tool change position and then the X and Y, simultaneously. This is the safest choice, in most cases; however, if no clearance issues are evident, then simultaneous movement of X, Y and Z may be chosen. Note that this movement is always at rapid traverse.

What type of coordinate system should be used, Multi Mode?

The three choices for establishing a work coordinate system are: Workpiece Coordinate (WPC), Multi 5 X 2 and Offset type.

WPC

With this selection set to OFF, the ordinary Work Piece Coordinate (WPC) system applies. Values are set for the location of the origin of the coordinate system, just as with G-Code programs, additionally offsets G54–59 may be used.

Part 6   Conversational Programming

Multi 5 X 2

By selecting Multi 5 X 2, the machining of multiple repetitions of the same program for several workpieces can be used. A Multi Flag is required, in conjunction with this call, in order to identify how many repetitions and where. This technique is limited to each of the repetitions having corresponding distances from part to part. For example, the distance between all repetitions in the X axis must be equal and the Y axis distances must be the same, as well. Up to ten duplications can be set.

Offset Type

Using this type of coordinate system arrangement allows the arbitrary location of the origin for multiple workpieces within the working envelope of the machine. The locations may be random and polar rotation of the coordinate system is allowed. Up to 10 individual offsets are allowed.

Coordinate System

In this unit the actual physical locations for the coordinate system axis zero points are entered. As mentioned earlier, this can be in the form of a WPC or work offset using G54-G59 as is common in G-Code programs. The operator measures the distances in X, Y and Z in relation to the Machine Zero and enters these values into the program in this unit.

Machining Unit

Here, the individual machining units are identified in order to complete the workpiece. In other words, what type of machining is to be done? In Figure 8 below note that the choices are: POINT MACHINING, LINE MACHINING, FACE MACHINING, MANUAL PROGRAM, OTHER, WPC, OFFSET, END and SHAPE CHECK. The following descriptions will be limited to Point machining and Line machining in the interest of space.

POINT MACH-ING	LINE MACH-ING	FACE MACH-ING	MANUAL PROGRAM	OTHER	WPC	OFFSET	END	SHAPE CHECK

Figure 8

Point Machining

Point machining constitutes a large percentage of Machining Center work. In the figure below, the choices for types of point machining are shown:

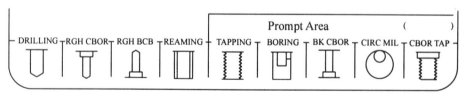

Figure 9

Part 6 Conversational Programming

When a selection is made from one of these choices, then unit data are required and automatic tool development is completed based on this information. The required information for Drilling is:

The diameter, the depth and whether the hole is to be chamfered. The basic required tooling is developed based on this information. For example, a center drill or spot drill, a drill of the size stated and a chamfering cutter will be developed. This tooling information is taken from a predetermined tool file in the control. The tool file should be constructed (for all other types of machining units) by the machinist/programmer prior to programming but can be done as the program is completed.

Sequence Data

Tool Sequence Data

Each individual tool has specific sequence data that are required as follows:

- definition of the actual size of the tool
- the priority in which this tool is to be used
- the diameter of the hole
- the hole depth
- pre-existing hole diameter
- pre-existing hole depth
- the desired surface finish
- the type of drilling cycle (i.e. drilling, pecking etc.)
- the cutting speeds and feeds and, the use of any M-Codes.

Shape Sequence Data

Finally, in the machining unit is found the shape sequence data. The actual figure pattern or shape is identified. In the case of drilling the choices are: POINT, LINE, SQUARE, GRID, CIRCLE, ARC and CHORD, shown in Figure 10. As soon as the pattern is completed, SHAPE END is pressed to end the unit.

Just as with all units, CHECK allows the shape to be checked graphically for each individual unit. The remainder of the program is constructed in the same manner until all geometry shapes are complete.

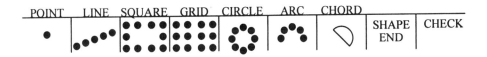

Figure 10

Part 6 Conversational Programming

Line Machining

Line machining (linear contouring) is another very common activity performed by Machining Centers. In the figure below, choices for types of line machining are shown.

Figure 11

When a selection is made, unit data are required and automatic tool development is completed, based on this information. The required information for line machining is: the depth of cut; the amount of stock removal in Z (SRV-Z), the amount of radial stock removal in X-Y (SRV-R); the desired finished surface roughness; chamfer width, if required; the allowance for the finish Z depth cut; and, the allowance for the radial finish width cut. The values input to these items determine the automatic tool development.

Tool Sequence Data

Each individual tool developed has specific sequence data that are required as follows: the nominal diameter of the tool; the priority in which the tool is to be used; the approach point along the X axis, the approach point along the Y axis, the cutting direction of either CW or CCW, the plunge cutting feedrate along the Z axis; the depth of cut, the cutting speed, the cutting feedrate; and, any M-Codes required.

Shape Sequence Data

Finally, the shape sequence data is created in the machining unit where the actual figure pattern or shape is identified. In the case of line machining, the choices are: SQUARE, CIRCLE, and ARBITRARY, shown in the Figure 12.

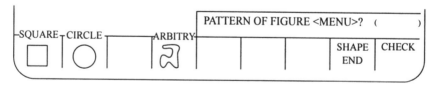

Figure 12

The geometric shape is constructed by input of point data that describe each feature of square, circular or arbitrary shape. As soon as the pattern is completed, SHAPE END is pressed to end the unit.

Just as with all similar units, CHECK allows the shape to be checked graphically for each individual unit. The remainder of the program is constructed in the same manner until all geometry shapes are complete.

Part 6   Conversational Programming

End Unit

This unit ends the program in the same manner as M30 does in a G-Code program. The programmer has the option in this unit of continuing the program for a number of repetitions and to control the Automatic Tool Change positioning.

Shape Check

When the entire program is written, it is beneficial to verify its accuracy by performing first a Shape Check and then Tool Path Check. The shape check verifies the geometry and the tool path verifies the actual relationship between the geometry and the tools actual cutting path. The machinist/programmer has the ability to change from two-dimensional to three-dimensional views or split the display screen to show the X-Y and X-Z and zoom-in on features that are hard to see. Once these checks are completed without any errors, the program is ready for set-up of the tool and work offsets. Then the machinist may begin automatic operation.

The Future of CNC Programming

The ultimate goal of any manufacturing is to increase productivity by improving efficiency. Minimizing wasted and idle time and to cut lead times while maintaining accuracy. Modern technology enables these goals to become reality.

The PC has revolutionized our society and manufacturing has been a benefactor because of the direct effect on controlling CNC machines. Now PC's are used to network machines together, in order to manage workloads. These networks also make it possible to communicate between machines and the office to manage and download programs, to obtain machine status and operation reports in real time, and monitor other network locations. Machine monitoring

Figure 13

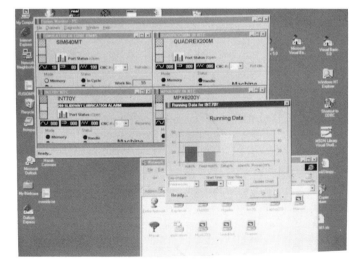

Figure 14

Part 6   Conversational Programming

includes automatic operation, machine stop and feed hold, set-up and alarm. Spindle load and spindle speed are recorded to provide reports and information on completed workpiece counts. Some machines utilize both a PC and CNC fused into one, providing bi-directional communication between the PC and CNC. This makes an intelligent CNC control system. Intelligent CNC responds to questions, makes suggestions and provides detailed reports on machine operation and production status.

The most advanced RISC CPU (Reduced Instruction Set Computer, Central Processing Unit) technology is used providing faster processing speeds which, in turn, help to achieve reduced set-up and cycle times.

Knowledge-based navigation functions allow the determination of optimum metal cutting conditions prior to actual cutting, dependant upon stored data. Based on the part program, tool data and workpiece material, the navigation functions suggest the optimum cutting conditions and shows where improvements in cycle time can be achieved through changes in spindle speed, feed rate and tools.

Advancements in the graphical cutting simulation allow the 3-dimensional solid part model to be displayed. This feature can be used to show part sections for checking inside diameters and deep holes, and views can be rotated. These added capabilities aid in program verification of tool paths.

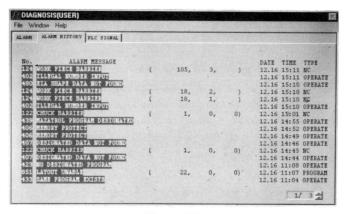

Figure 15

To keep machine utilization high, modern controls allow self-diagnostic functions for service and maintenance. Self-diagnostic menus quickly trouble shoot the cause of an alarm and suggest possible solutions for trouble-shooting. Alarm displays indicate when scheduled maintenance is required and on-line support is available.

As technological advancements continue at an unbelievable pace the manufacturing industry will undoubtedly benefit. Innovative new methods for CAD/CAM and Conversational Programming of CNC machines will continue to emerge. For the machinist/programmer this means that the new methods and tools used for programming will require a life-long learning approach. This book has been intended to begin that learning process.

APPENDIX

APPENDIX

Chart 1

ENGLISH DRILL SIZES

Drill/Decimal		Drill/Decimal		Drill/Decimal		Drill/Decimal	
80	.0135	40	.0980	2	.2210	33/64	.5156
79	.0145	39	.0995	1	.2280	17/32	.5312
1/64	.0156	38	.1015	A	.2340	35/64	.5469
78	.0160	37	.1040	15/64	.2344	**9/16**	**.5625**
77	.0180	36	.1065	B	.2380	37/64	.5781
76	.0200	7/64	.1094	C	.2420	19/32	.5938
75	.0210	35	.1100	D	.2460	39/64	.6094
74	.0225	34	.1110	E	.2500	**5/8**	**.6250**
73	.0240	33	.1130	**1/4**	**.2500**	41/64	.6406
72	.0250	32	.1160	F	.2570	21/32	.6562
71	.0260	31	.1200	G	.2610	43/64	.6719
70	.0280	**1/8**	**.1250**	17/64	.2656	**11/16**	**.6875**
69	.0292	30	.1285	H	.2660	45/64	.7031
68	.0310	29	.1360	I	.2720	23/32	.7188
1/32	.0312	28	.1405	J	.2770	47/64	.7344
67	.0320	9/64	.1406	K	.2810	**3/4**	**.7500**
66	.0330	27	.1440	9/32	.2812	49/64	.7656
65	.0350	26	.1470	L	.2900	25/32	.7812
64	.0360	25	.1495	M	.2950	51/64	.7969
63	.0370	24	.1520	19/64	.2969	**13/16**	**.8125**
62	.0380	23	.1540	N	.3020	53/64	.8281
61	.0390	5/32	.1562	**5/16**	**.3125**	27/32	.8438
60	.0400	22	.1570	O	.3160	55/64	.8594
59	.0410	21	.1590	P	.3230	**7/8**	**.8750**
58	.0420	20	.1610	21/64	.3281	57/64	.8906
57	.0430	19	.1660	Q	.3320	29/32	.9062
56	.0465	18	.1695	R	.3390	59/64	.9219
3/64	.0469	11/64	.1719	11/32	.3438	**15/16**	**.9375**
55	.0520	17	.1730	S	.3480	61/64	.9531
54	.0550	16	.1770	T	.3580	31/32	.9688
53	.0595	15	.1800	23/64	.3594	63/64	.9844
1/16	**.0625**	14	.1820	U	.3680	1	1.0000
52	.0635	13	.1850	**3/8**	**.3750**	+64th increments up to 1-7/8"	
51	.0670	**3/16**	**.1875**	V	.3770		
50	.0700	12	.1890	W	.3860		
49	.0730	11	.1910	25/64	.3906		
48	.0760	10	.1935	X	.3970		
5/64	.0780	19	.1960	Y	.4040	+32nd increments up to 2-1/4"	
47	.0785	8	.1990	13/32	.4062		
46	.0810	7	.2010	Z	.4130		
45	.0820	13/64	.2031	27/64	.4219		
44	.0860	6	.2040	**7/16**	**.4375**		
43	.0890	5	.2055	29/64	.4531	+16th increments up to 4-1/4"	
42	.0935	4	.2090	15/32	.4688		
3/32	.0938	3	.2130	31/64	.4844		
41	.0960	7/32	.2188	**1/2**	**.5000**		

Appendix

Chart 2

METRIC THREADS

Basic Thread Size*	Drill Size**
M1.6x0.35	1.25mm or #55
M2x0.4	1.60mm or #52
M2.5x0.45	2.05mm or #46
M3x0.5	2.50mm or #39
M3.5x0.6	2.90mm or #32
M4x0.7	3.30mm or #30
M5x0.8	4.20mm or #19
M6x1	5.00mm or #8
M8x1.25	6.80mm or H
M8x1	7.00mm or J
M10x1.5	8.50mm or R
M10x1.25	8.80mm or 11/32
M12x1.75	10.20mm or 3/32
M12x1.25	10.80mm or 7/64
M14x2	12.00mm or 5/32
M14x1.5	12.50mm or 1/2
M16x2	14.00mm or 5/64
M16x1.5	14.50mm or 7/64
M18x2.5	15.50mm or 9/64
M18x1.5	16.50mm or 1/32
M20x2.5	17.50mm or 1/16
M20x1.5	18.50mm or 7/64
M22x2.5	19.50mm or 9/64
M22x1.5	20.50mm or 13/16
M24x3	21.00mm or 53/64
M24x2	22.00mm or 7/8
M27x3	24.00mm or 15/16
M27x2	25.00mm or 1

* Pitch callout not required on metric coarse series.
** Closest size for 75% theoretical thread.

Appendix

Chart 3

ENGLISH DRILL SIZES

Drill/Decimal		Drill/Decimal		Drill/Decimal		Drill/Decimal	
80	.0135	40	.0980	2	.2210	33/64	.5156
79	.0145	39	.0995	1	.2280	17/32	.5312
1/64	.0156	38	.1015	A	.2340	35/64	.5469
78	.0160	37	.1040	15/64	.2344	**9/16**	**.5625**
77	.0180	36	.1065	B	.2380	37/64	.5781
76	.0200	7/64	.1094	C	.2420	19/32	.5938
75	.0210	35	.1100	D	.2460	39/64	.6094
74	.0225	34	.1110	E	.2500	**5/8**	**.6250**
73	.0240	33	.1130	**1/4**	**.2500**	41/64	.6406
72	.0250	32	.1160	F	.2570	21/32	.6562
71	.0260	31	.1200	G	.2610	43/64	.6719
70	.0280	**1/8**	**.1250**	17/64	.2656	**11/16**	**.6875**
69	.0292	30	.1285	H	.2660	45/64	.7031
68	.0310	29	.1360	I	.2720	23/32	.7188
1/32	.0312	28	.1405	J	.2770	47/64	.7344
67	.0320	9/64	.1406	K	.2810	**3/4**	**.7500**
66	.0330	27	.1440	9/32	.2812	49/64	.7656
65	.0350	26	.1470	L	.2900	25/32	.7812
64	.0360	25	.1495	M	.2950	51/64	.7969
63	.0370	24	.1520	19/64	.2969	**13/16**	**.8125**
62	.0380	23	.1540	N	.3020	53/64	.8281
61	.0390	5/32	.1562	**5/16**	**.3125**	27/32	.8438
60	.0400	22	.1570	O	.3160	55/64	.8594
59	.0410	21	.1590	P	.3230	**7/8**	**.8750**
58	.0420	20	.1610	21/64	.3281	57/64	.8906
57	.0430	19	.1660	Q	.3320	29/32	.9062
56	.0465	18	.1695	R	.3390	59/64	.9219
3/64	.0469	11/64	.1719	11/32	.3438	**15/16**	**.9375**
55	.0520	17	.1730	S	.3480	61/64	.9531
54	.0550	16	.1770	T	.3580	31/32	.9688
53	.0595	15	.1800	23/64	.3594	63/64	.9844
1/16	.0625	14	.1820	U	.3680	1	1.0000
52	.0635	13	.1850	3/8	.3750	+64th increments up to 1-7/8"	
51	.0670	3/16	.1875	V	.3770		
50	.0700	12	.1890	W	.3860		
49	.0730	11	.1910	25/64	.3906		
48	.0760	10	.1935	X	.3970		
5/64	.0780	19	.1960	Y	.4040	+32nd increments up to 2-1/4"	
47	.0785	8	.1990	13/32	.4062		
46	.0810	7	.2010	Z	.4130		
45	.0820	13/64	.2031	27/64	.4219		
44	.0860	6	.2040	**7/16**	**.4375**		
43	.0890	5	.2055	29/64	.4531	+16th increments up to 4-1/4"	
42	.0935	4	.2090	15/32	.4688		
3/32	.0938	3	.2130	31/64	.4844		
41	.0960	7/32	.2188	**1/2**	**.5000**		

Appendix

Chart 4
METRIC DRILL SIZES

Drill/Decimal		Drill/Decimal		Drill/Decimal		Drill/Decimal	
.35mm	.0138	2.50mm	.0984	6.20mm	.2441	**10.00mm**	**.3937**
.38mm	.0150	2.55mm	.1004	6.25mm	.2461	10.20mm	.4016
.40mm	.0157	2.60mm	.1024	6.30mm	.2480	10.50mm	.4134
.42mm	.0165	2.65mm	.1043	6.40mm	.2520	10.80mm	.4252
.45mm	.0177	2.70mm	.1063	6.50mm	.2559	**11.00mm**	**.4331**
.48mm	.0189	2.75mm	.1083	6.60mm	.2598	11.20mm	.4409
.50mm	.0197	2.80mm	.1102	6.70mm	.2638	11.50mm	.4528
.55mm	.0217	2.90mm	.1142	6.75mm	.2657	11.80mm	.4646
.60mm	.0236	**3.00mm**	**.1181**	6.80mm	.2677	**12.00mm**	**.4724**
.65mm	.0256	3.10mm	.1220	6.90mm	.2717	12.20mm	.4803
.70mm	.0276	3.20mm	.1260	**7.00mm**	**.2756**	12.50mm	.4921
.75mm	.0295	3.25mm	.1280	7.10mm	.2795	**13.00mm**	**.5118**
.80mm	.0315	3.30mm	.1299	7.20mm	.2835	13.50mm	.5315
.85mm	.0335	3.40mm	.1339	7.25mm	.2854	**14.00mm**	**.5512**
.90mm	.0354	3.50mm	.1378	7.30mm	.2874	14.50mm	.5709
.95mm	.0374	3.60mm	.1417	7.40mm	.2913	**15.00mm**	**.5906**
1.00mm	**.0394**	3.70mm	.1457	7.50mm	.2953	15.50mm	.6102
1.05mm	.0413	3.75mm	.1476	7.60mm	.2992	**16.00mm**	**.6299**
1.10mm	.0433	3.80mm	.1496	7.70mm	.3031	16.50mm	.6496
1.15mm	.0453	3.90mm	.1535	7.75mm	.3051	**17.00mm**	**.6693**
1.20mm	.0472	**4.00mm**	**.1575**	7.80mm	.3071	17.50mm	.6890
1.25mm	.0492	4.10mm	.1614	7.90mm	.3110	**18.00mm**	**.7087**
1.30mm	.0512	4.20mm	.1654	**8.00mm**	**.3150**	18.50mm	.7283
1.35mm	.0531	4.25mm	.1673	8.10mm	.3189	**19.00mm**	**.7480**
1.40mm	.0551	4.30mm	.1693	8.20mm	.3228	19.50mm	.7677
1.45mm	.0571	4.40mm	.1732	8.25mm	.3248	**20.00mm**	**.7874**
1.50mm	.0591	4.50mm	.1772	8.30mm	.3268	20.50mm	.8071
1.55mm	.0610	4.60mm	.1811	8.40mm	.3307	**21.00mm**	**.8268**
1.60mm	.0630	4.70mm	.1850	8.50mm	.3346	21.50mm	.8465
1.65mm	.0650	4.75mm	.1870	8.60mm	.3386	**22.00mm**	**.8661**
1.70mm	.0669	4.80mm	.1890	8.70mm	.3425	22.50mm	.8858
1.75mm	.0689	4.90mm	.1929	8.75mm	.3445	**23.00mm**	**.9055**
1.80mm	.0700	**5.00mm**	**.1969**	8.80mm	.3465	23.50mm	.9252
1.85mm	.0728	5.10mm	.2008	8.90mm	.3504	**24.00mm**	**.9449**
1.90mm	.0748	5.20mm	.2047	**9.00mm**	**.3543**	24.50mm	.9646
1.95mm	.0768	5.25mm	.2067	9.10mm	.3583	**25.00mm**	**.9843**
2.00mm	**.0787**	5.30mm	.2087	9.20mm	.3622	+1.00mm increments up to 48mm	
2.05mm	.0807	5.40mm	.2126	9.25mm	.3642		
2.10mm	.0827	5.50mm	.2165	9.30mm	.3661		
2.15mm	.0846	5.60mm	.2205	9.40mm	.3701		
2.20mm	.0866	5.70mm	.2244	9.50mm	.3740		
2.25mm	.0886	5.75mm	.2264	9.60mm	.3780	+5.00mm increments from 50mm up to 105mm	
2.30mm	.0906	5.80mm	.2283	9.70mm	.3819		
2.35mm	.0925	5.90mm	.2323	9.75mm	.3839		
2.40mm	.0945	**6.00mm**	**.2362**	9.80mm	.3858		
2.45mm	.0965	6.10mm	.2402	9.90mm	.3898		

Appendix
Chart 5

COUNTERBORED HOLES

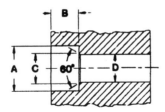

USA SOCKET-HEAD CAP SCREWS

SCREW DIA	A COUNTERBORE DIA	B COUNTERBORE DEPTH	C COUNTERSINK DIA	D CLEARANCE DIA	
				NORMAL FIT	CLOSE FIT
#0	1/8	.060	.074	#49	#51
#2	3/16	.086	.102	#36	3/32
#4	7/32	.112	.130	#29	1/8
#5	1/4	.125	.145	#23	9/64
#6	9/32	.138	.158	#18	#23
#8	5/16	.164	.188	#10	#15
#10	3/8	.190	.218	#2	#5
1/4	7/16	.250	.278	9/32	17/64
5/16	17/32	.312	.346	11/32	21/64
3/8	5/8	.375	.415	13/32	25/64
7/16	23/32	.438	.483	15/32	29/64
1/2	13/16	.500	.552	17/32	33/64
5/8	1	.625	.689	21/32	41/64
3/4	1-3/16	.750	.828	25/32	49/64
7/8	1-3/8	.875	.963	29/32	57/64
1	1-5/8	1.000	1.100	1-1/32	1-1/64
1-1/4	2	1.250	1.370	1-5/16	1-9/32
1-1/2	2-3/8	1.500	1.640	1-9/16	1-17/32
1-3/4	2-3/4	1.750	1.910	1-13/16	1-25/32
2	3-1/8	2.000	2.180	2-1/16	2-1/32

Appendix
Chart 6

COUNTERBORED HOLES

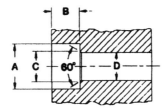

METRIC SOCKET-HEAD CAP SCREWS

SCREW DIA	A COUNTERBORE DIA	B COUNTERBORE DEPTH	C COUNTERSINK DIA	D CLEARANCE DIA NORMAL FIT	D CLEARANCE DIA CLOSE FIT
M1.6	3.50mm	1.6mm	2.0mm	1.95mm	1.80mm
M2	4.40mm	2mm	2.6mm	2.40mm	2.20mm
M2.5	5.40mm	2.5mm	3.1mm	3.00mm	2.70mm
M3	6.50mm	3mm	3.6mm	3.70mm	3.40mm
M4	8.25mm	4mm	4.7mm	4.80mm	4.40mm
M5	9.75mm	5mm	5.7mm	5.80mm	5.40mm
M6	11.20mm	6mm	6.8mm	6.80mm	6.40mm
M8	14.50mm	8mm	9.2mm	8.80mm	8.40mm
M10	17.50mm	10mm	11.2mm	10.80mm	10.50mm
M12	19.50mm	12mm	14.2mm	13.00mm	12.50mm
M14	22.50mm	14mm	16.2mm	15.00mm	14.50mm
M16	25.50mm	16mm	18.2mm	17.00mm	16.50mm
M20	31.50mm	20mm	22.4mm	21.00mm	20.50mm
M24	37.50mm	24mm	26.4mm	25.00mm	24.50mm
M30	47.50mm	30mm	33.4mm	31.50mm	31.00mm
M36	56.50mm	36mm	39.4mm	37.50mm	37.00mm
M42	66.00mm	42mm	45.6mm	44.00mm	43.00mm
M48	75.00mm	48mm	52.6mm	50.00mm	49.00mm

Appendix

Chart 7

GEOMETRIC SYMBOLS AND DEFINITIONS

▱ .002	⊕ ⌀.005 Ⓜ A

Feature-Control Frame
A specification box that shows a particular geometric characteristic (flatness, straightness, etc.) applied to a part feature and states the allowable tolerance. The feature's tolerance may be individual, or related to one or more datums. Any datum references and tolerance modifiers are also shown.

—A—	A1 A2 A3
Datum Feature	**Datum Targets**
A flag which designates a physical feature of the part to be used as a reference to measure geometric characteristics of other part features.	Callouts occasionally needed to designate specific points, lines, or areas on an actual part to be used to establish a theoretical datum feature.
1.500	⌀
Basic Dimension	**Cylindrical Tolerance Zone**
A box around any drawing dimension makes it a "basic" dimension, a theoretically exact value used as a reference for measuring geometric characteristics and tolerances of other part features.	This symbol, commonly used to indicate a diameter dimension, also specifies a cylindrically shaped tolerance zone in a feature-control frame.
Ⓜ	Ⓛ
Maximum Material Condition	**Least Material Condition**
Abbreviation: MMC. A tolerance modifier that applies the stated tight tolerance zone only while the part theoretically contains the maximum amount of material permitted within its dimensional limits (e.g. minimum hole diameters and maximum shaft diameters), allowing more variation under normal conditions.	Abbreviation: LMC. A tolerance modifier that applies the stated tight tolerance zone only while the part theoretically contains the minimum amount of material permitted within its dimensional limits (e.g. maximum hole diameters and minimum shaft diameters), allowing more variation under normal conditions.
Ⓢ	.500 Ⓟ
Regardless of Feature Size	**Projected Tolerance Zone**
Abbreviation: RFS. A tolerance modifier that applies the stated tight tolerance zone under all size conditions. RFS is generally assumed if neither MMC nor LMC are stated.	An additional specification box attached underneath a feature-control frame. It extends the feature's tolerance zone beyond the part's surface by the stated distance, ensuring perpendicularity for proper alignment of mating parts.

Appendix

Chart 8
GEOMETRIC CHARACTERIDSTICS

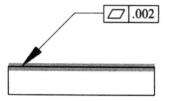

⌗ **Flatness**

All points on the indicated surface must lie in a single plane, within the specified tolerance zone.

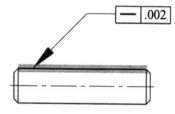

— **Straightness**

All points on the indicated surface or axis must lie in a straight line in the direction shown, within the specified tolerance zone

---

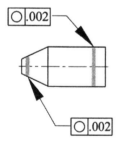

○ **Circularity (Roundness)**

If the indicated surface were sliced by any plane perpendicular to its axis, the resulting outline must be a perfect circle, within the specified tolerance zone.

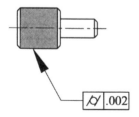

⌭ **Cylindricity**

All points on the indicated surface must lie in a perfect cylinder around a center axis, within the specified tolerance zone.

Appendix

Chart 9
GEOMETRIC CHARACTERIDSTICS

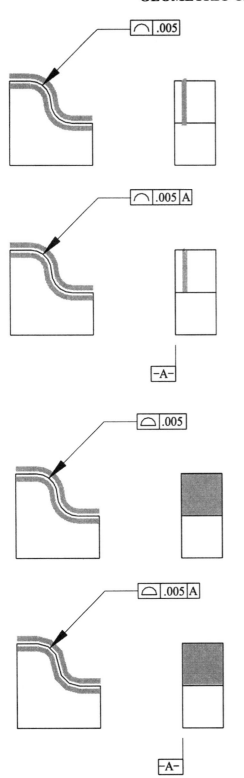

⌒ **Linear Profile**

All points on any full slice of the indicated surface must lie on its theoretical two-dimensional profile, as defined by basic dimensions, within the specified tolerance zone. The profile may or may not be oriented with respect to daatums.

⌓ **Surface Profile**

All points on the indicated surface must lie on its theoretical three-dimensional profile, as defined by basic dimensions, within the specified tolerance zone. The profile may or may not be oriented with respect to datums.

Appendix

Chart 10
GEOMETRIC CHARACTERIDSTICS

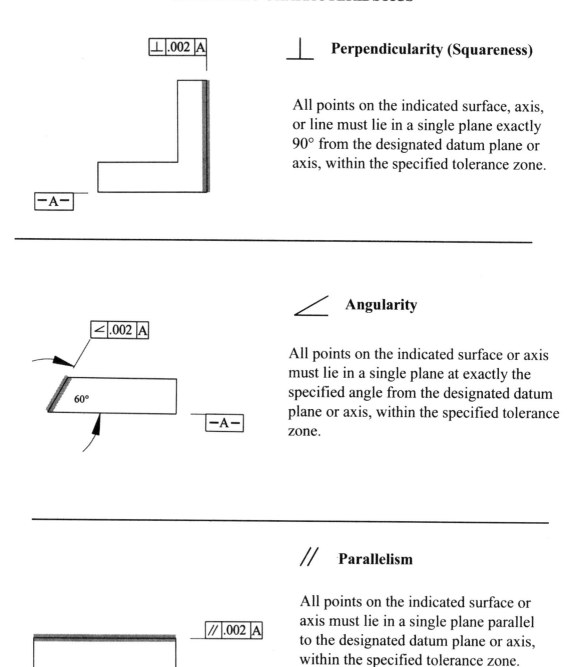

⊥ **Perpendicularity (Squareness)**

All points on the indicated surface, axis, or line must lie in a single plane exactly 90° from the designated datum plane or axis, within the specified tolerance zone.

∠ **Angularity**

All points on the indicated surface or axis must lie in a single plane at exactly the specified angle from the designated datum plane or axis, within the specified tolerance zone.

// **Parallelism**

All points on the indicated surface or axis must lie in a single plane parallel to the designated datum plane or axis, within the specified tolerance zone.

Appendix

Chart 11
GEOMETRIC CHARACTERIDSTICS

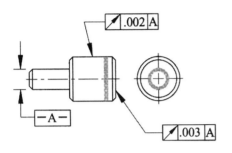

✒ Circular Runout

Each circular element of the indicated surface is allowed to deviate only the specified amount from its theoretical form and orientation during 360° rotation about the designate datum axis.

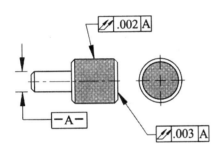

✒✒ Total Runout

The entire indicated surface is allowed to deviate only the specified amount from its theoretical form and orientation during 360° rotation about the designated datum axis.

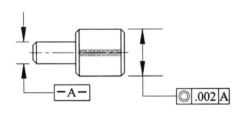

◎ Concentricity

If the indicated surface were sliced by any plane perpendicular to the designated datum axis, every slice's center of area must lie on the datum axis. within the specified cylindrical tolerance zone (controls rotational balance).

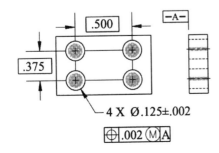

⊕ Position (Replaces ═ Symetry)

The indicated feature's axis must be located within the specified tolerance zone from its true theoretical position, correctly oriented relative to the designated datum plane or axis.

Appendix

BASIC MATHEMATICAL FORMULAS

Chart 12

TRIGONOMETRY FUNCTIONS

SINE

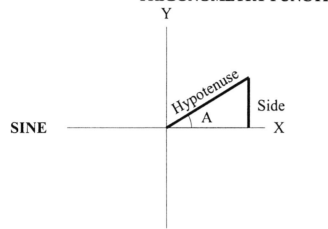

$$\sin A = \frac{\text{Side}}{\text{Hypotenuse}}$$

COSINE

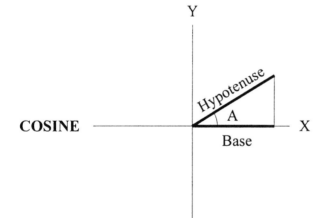

$$\cos A = \frac{\text{Base}}{\text{Hypotenuse}}$$

TANGENT

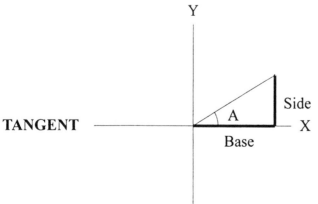

$$\tan A = \frac{\text{Side}}{\text{Base}}$$

Appendix

Chart 13

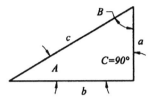

RIGHT TRIANGLES

Known Sides and Angles	Unknown Sides and Angles			Area
a and b	$c = \sqrt{a^2 + b^2}$	$A = \arctan \dfrac{a}{b}$	$B = \arctan \dfrac{b}{a}$	$\dfrac{a \times b}{2}$
a and c	$b = \sqrt{c^2 - a^2}$	$A = \arcsin \dfrac{a}{c}$	$A = \arccos \dfrac{a}{c}$	$\dfrac{a \times \sqrt{c^2 - a^2}}{2}$
b and c	$a = \sqrt{c^2 - b^2}$	$A = \arccos \dfrac{b}{c}$	$B = \arcsin \dfrac{b}{c}$	$\dfrac{b \times \sqrt{c^2 - b^2}}{2}$
a and $\angle A$	$b = \dfrac{a}{\tan A}$	$c = \dfrac{a}{\sin A}$	$B = 90° - A$	$\dfrac{a^2}{2 \times \tan A}$
a and $\angle B$	$b = a \times \tan B$	$c = \dfrac{a}{\cos B}$	$A = 90° - B$	$\dfrac{a^2 \times \tan B}{2}$
b and $\angle A$	$a = b \times \tan A$	$c = \dfrac{b}{\cos A}$	$B = 90° - A$	$\dfrac{b^2 \times \tan A}{2}$
b and $\angle B$	$a = \dfrac{b}{\tan B}$	$c = \dfrac{b}{\sin B}$	$A = 90° - B$	$\dfrac{b^2}{2 \times \tan B}$
c and $\angle A$	$a = c \times \sin A$	$b = c \times \cos A$	$B = 90° - A$	$c^2 \times \sin A \times \cos$
c and $\angle B$	$a = c \times \cos B$	$b = c \times \sin B$	$A = 90° - B$	$c^2 \times \sin B \times \cos$

Appendix

Chart 14

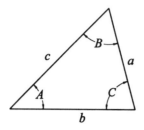

OBLIQUE TRIANGLES

Known Sides and Angles	Unknown Sides and Angles			Area
All three sides a, b, c	$A = \arccos \dfrac{b^2 + c^2 - a^2}{2bc}$	$B = \arcsin \dfrac{b \times \sin A}{a}$	$C = 180° - A - B$	$\dfrac{a \times b \times \sin C}{2}$
Two sides and the angle between them $a, b, \angle C$	$c = \sqrt{a^2 + b^2 - (2ab \times \cos C)}$	$A = \arctan \dfrac{a \times \sin C}{b - (a \times \cos C)}$	$B = 180° - A - C$	$\dfrac{a \times b \times \sin C}{2}$
Two sides and the angle opposite one of the sides $a, b, \angle A$ ($\angle B$ less than $90°$)	$B = \arcsin \dfrac{b \times \sin A}{a}$	$C = 180° - A - B$	$c = \dfrac{a \times \sin C}{\sin A}$	$\dfrac{a \times b \times \sin C}{2}$
Two sides and the angle opposite one of the sides $a, b, \angle A$ ($\angle B$ greater than $90°$)	$B = 180° - \arcsin \dfrac{b \times \sin A}{a}$	$C = 180° - A - B$	$c = \dfrac{a \times \sin C}{\sin A}$	$\dfrac{a \times b \times \sin C}{2}$
One side and two angles $a, \angle A, \angle B$	$b = \dfrac{a \times \sin B}{\sin A}$	$C = 180° - A - B$	$c = \dfrac{a \times \sin C}{\sin A}$	$\dfrac{a \times b \times \sin C}{2}$

Appendix

BASIC MATHEMATICAL FORMULAS USED IN PROGRAMMING TURNING CENTERS WHEN THE CONTROL PANEL IS UTILIZED TO PROGRAM THE MACHINE

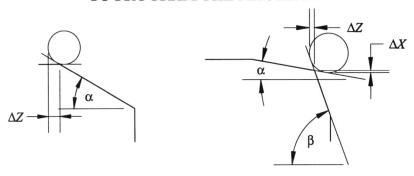

Figure 14   **Figure 15**

$$\Delta X = r\left[1 - \tan\left(45 - \frac{\alpha}{2}\right)\right] \quad \Delta Z = r\left(1 - \tan\frac{\alpha}{2}\right)$$

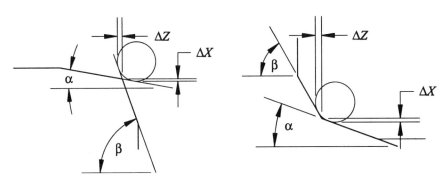

Figure 16   **Figure 17**

$$\Delta X = r\left(1 - \frac{\cos[(\alpha + \beta)/2]}{\cos[(\alpha - \beta)/2]}\right) \quad \Delta Z = r\left(\frac{\cos[(\alpha + \beta)/2]}{\cos[(\alpha - \beta)/2]}\right)$$

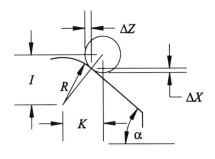 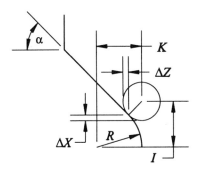

Figure 18   **Figure 19**

$$\Delta X = r(1 - \cos\alpha) \quad I = (R + r)\cos\alpha$$
$$\Delta Z = r(1 - \sin\alpha) \quad K = (R + r)\sin\alpha$$

Appendix

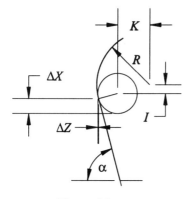

Figure 20

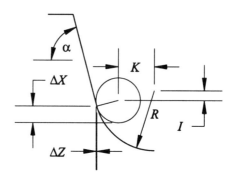

Figure 21

$$\Delta X = r(1 - \cos \alpha) \quad I = (R - r)\cos \alpha$$
$$\Delta Z = r(1 - \sin \alpha) \quad K = (R - r)\sin \alpha$$

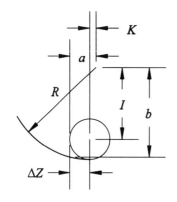

Figure 22

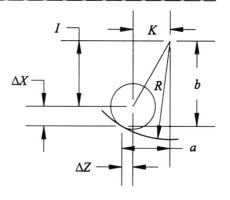

Figure 23

$$\Delta Z = r$$
$$\Delta X = r - [b - (R - r)^2 - (a - r)^2]$$
$$I = (R - r)^2 - (a - r)^2$$
$$K = a - r$$
$$\Delta X = r$$
$$\Delta Z = r - [a - (R - r)^2 - (b - r)^2]$$
$$I = b - r$$
$$K = (R - r)^2 - (b - r)^2$$

Appendix

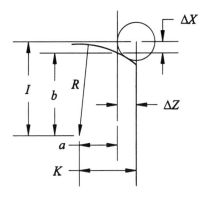

Figure 24　　　　　　　**Figure 25**

$$\Delta Z = r$$
$$\Delta X = r - [(R + r)^2 - (a + r)^2 - b]$$
$$I = (R + r)^2 - (a + r)^2$$
$$K = a + r$$
$$\Delta X = r$$
$$\Delta Z = r - [(R + r)^2 - (b + r)^2 - a]$$
$$I = b + r$$
$$K = (R + r)^2 - (b + r)^2$$

BASIC MATHEMATICAL FORMULAS USED IN PROGRAMMING MACHINING CENTERS WHEN THE CONTROL PANEL IS UTILIZED TO PROGRAM THE MACHINE

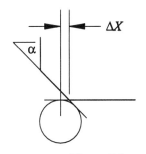
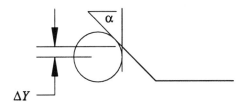

Figure 26　　　　　　　**Figure 27**

$$\Delta Y = r \times \tan\left(45 - \frac{\alpha}{2}\right) \quad \Delta X = r \times \tan\frac{\alpha}{2}$$

Appendix

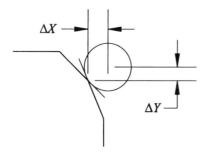

Figure 28

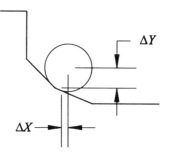

Figure 29

$$\Delta Y = r \times \frac{\cos[(\alpha + \beta)/2]}{\cos[(\alpha - \beta)/2]} \quad \Delta X = r \times \frac{\cos[(\alpha + \beta)/2]}{\cos[(\alpha - \beta)/2]}$$

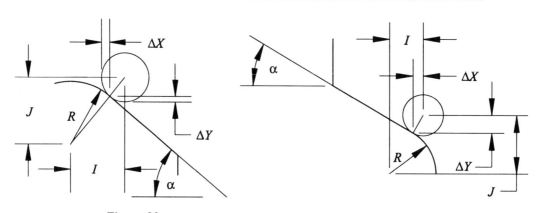

Figure 30

Figure 31

$$\Delta X = r \times \sin \alpha \quad I = (R + r) \times \sin \alpha$$
$$\Delta Y = r \times \cos \alpha \quad J = (R + r) \times \cos \alpha$$

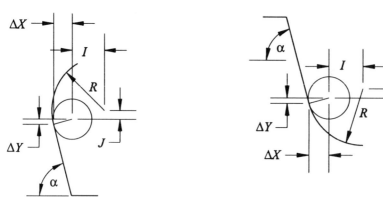

Figure 32

Figure 33

$$\Delta X = r \times \sin \alpha \quad I = (R - r) \times \sin \alpha$$
$$\Delta Y = r \times \cos \alpha \quad J = (R - r) \times \cos \alpha$$

Appendix

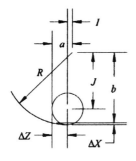

Appendix Figure 34

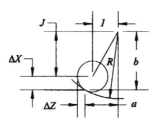

Appendix Figure 35

$\Delta X = r$
$\Delta Y = b - \sqrt{(R-r)^2 - (a-r)^2}$
$I = a - r$
$J = b - \Delta Y$

$\Delta X = a - \sqrt{(R-r)^2 - (b-r)^2}$
$\Delta Y = r$
$I = a - \Delta X$
$J = b - r$

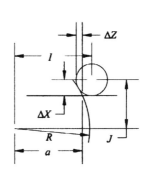

Appendix Figure 36

Appendix Figure 37

$\Delta X = r$
$\Delta Y = \sqrt{(R+r)^2 - (a+r)^2} - b$
$I = a + r$
$J = \sqrt{(R+r)^2 - (a+r)^2}$

$\Delta X = \sqrt{(R+r)^2 - (b+r)^2} - a$
$\Delta Y = r$
$I = \sqrt{(R+r)^2 - (b+r)^2}$
$J = b + r$

Appendix

Glossary

Popular Acronyms

AOS	Algebraic Order System
APT	Automatically Programmed Tools
ATC	Automatic Tool Changer
ANSI	American National Standards Institute
ASCII	American Standard Code for Information Interchange
ASME	American Society of Mechanical Engineers
BCD	Binary Coded Decimal
CAD	Computer Aided Design
CAM	Computer Aided Manufacturing
CAD/CAM	Computer Aided Design & Computer Aided Manufacturing
CIM	Computer-Integrated Manufacturing
CW	Clockwise
CCW	Counterclockwise
CD-ROM	Compact Disc-Read Only Memory
CD-RW	Compact Disc-Re-Writable
CDC	Cutter Diameter Compensation
CRC	Cutter Radius Compensation
CNC	Computer Numerical Control
CMM	Coordinate Measuring Machine
CPU	Central Processing Unit
CRT	Cathode Ray Tube
DNC	Direct Numerical Control
DVD	Digital Video Disc
DXF	Drawing Exchange Format
EDM	Electronic Discharge Machine
EIA	Electronics Industries Association
EOB	End of Block
FMS	Flexible Manufacturing System
GB	Gigabit
G-Code	Preparatory Functions (commands)

Glossary

GD&T	Geometric Dimensioning & Tolerancing	
GHz	Gigahertz	
GUI	Graphical User Interface	
HP	Horsepower	
HSS	High Speed Steel	
ID	Inside Diameter	
IGES	Initial Graphics Exchange Specification	
IPM	Inches Per Minute (also in/min)	
IPR	Inches Per Revolution (also in/rev)	
ISO	International Standards Organization	
KW	Kilowatt	
LAN	Local Area Network	
LED	Liquid Emitting Diode	
LCD	Liquid Crystal Display	
MB	Megabit	
M-Codes	Miscellaneous Functions	
MCU	Machine Control Unit	
MDI	Manual Data Input	
MHz	Megahertz	
Mm	Millimeters	
NC	Numerical Control	
OD	Outside Diameter	
PC	Personal Computer	
PCMCI	Portable Computer Memory Card Interface	
PLC	Programmable Logic Controller	
PSI	Pounds Per Square Inch	
RAM	Random Access Memory	
RPM	Revolutions Per Minute (also rev/min or r/min)	
RISC	Reduced Instruction Set Computer	
ROM	Read Only Memory	
R-8	Taper Designation	
RS232	Industry Standard Cabling Interface	
SFM	Surface Feet per Minute	
SME	Society of Manufacturing Engineering	
TLO	Tool Length Offset	
TNRC	Tool Nose Radius Compensation	

Glossary

Definitions

A-Axis

The A-axis is an auxiliary rotary axis that rotates about the X axis. Angular movements are specified in decimal degrees. Positive angular values refer to counterclockwise rotation and negative angular values indicate clockwise rotation.

Absolute Dimension

All numerical values (dimensional measurements) are derived from a fixed origin or datum in the coordinate system.

Address

Commonly referred to as letter address, because in programming, each program word is preceded by a letter, in order to identify what function is to be executed. Examples of letter address are: S for spindle speed designation, T for tool identification, M for miscellaneous functions, G for preparatory functions, etc.

Auxiliary axis

An auxiliary axis is any axis that is in addition to the primary axes of X, Y, and Z. These axes can be rotary (A,B or C) or linear (U, V or W). They are also called secondary axes.

Axis

The axis is the primary identifier of the cutting tool direction of movement in relationship with the machine type and orientation. The three linear axes for a machining center are X, Y and Z, and they are perpendicular to each other. The rotary axes are; A,B and C.

B-Axis

The B-axis is an auxiliary rotary axis that rotates about the Y axis. Angular movements are specified in decimal degrees. Positive angular values refer to counterclockwise rotation and negative angular values indicate clockwise rotation.

Block

A single line of CNC code consisting of program words that identify what activities the machine is to execute. Generally, each block is preceded by a block or sequence number (N) and is followed by an "End of Block" (EOB) character, represented by the semicolon (;).

Block Skip

Block Skip is sometimes called Block Delete or Optional Block Skip. When the Block Delete, or Optional Block Skip button or switch is on, the controller skips execution of the program blocks that are preceded with "/" and end with the end

Glossary

of block (;) character. If the button or switch is off, the machine will execute the programmed blocks and disregards the "/" symbol.

C-Axis

The C-axis is an auxiliary rotary axis that rotates about the Z axis. Angular movements are specified in decimal degrees. Positive angular values refer to counterclockwise rotation and negative angular values clockwise.

Canned Cycle

The function of a given cycle is defined as a set of operations assigned to one block and performed automatically without any possibility of interruption. Examples of canned cycles are: Cycle G81, which will perform a simple drill cycle and G84, which will perform tapping. Canned Cycles require additional information such as coordinate locations, reference plane values, peck amounts, etc. Canned Cycles simplify the part program by decreasing programming time. Another name for Canned Cycles is Fixed Cycles.

Cartesian Coordinates

A coordinate system that consists of three axes (X, Y and Z) that are perpendicular to each other. A grid is formed consisting of numerical graduations, representing the distances from the intersection of the three axes called origin.

Circular Interpolation

A programming feature that enables programming of two axes simultaneously to create arcs and circles. Information generally needed is the location of the arc center, the arc radius, the starting and ending points of the arc and the direction of cutting motion.

Coordinates

Numerical values that define the positional location of points from a predetermined zero point or origin from within the Cartesian Coordinate System.

Datum

A datum is an exact point, axis, or plane. A datum is the origin from which either the dimensional location and/or the characteristics of features of a part are established.

Dry Run

Sometimes the CNC part program is executed with no part mounted, to verify the programmed path of the tool under automatic operation. The typical form of Dry run is set by activating the DRY RUN function on the control during automatic cycle, where all of the rapid and work feeds are changed to the rapid traverse feed set in the parameters instead of the programmed feed. DRY RUN is

Glossary

also used to check a new program on the machine without any work actually being performed by the tool. This is particularly useful on programs with long cycle times so the operator can progress through the program more quickly.

Dwell

Dwell is determined by the preparatory function G04 and by using the letter address P or X, which corresponds to the time duration of dwell (also U for lathes). When used, it causes a pause in the machining operation for the length of time indicated in seconds (X or U) milliseconds (P) or revolutions (depending on parameter setting).

End of Block Character

A special character represented by the semicolon (;) that identifies the end of a program block. Known by the abbreviation or acronym EOB or E-O-B.

End of program

A miscellaneous function (M30) is placed in the last line of a program to indicate the end of the part program. At this command, the spindle, coolant, and feed are stopped and the program is returned to its start.

F-Word

The F-Word is utilized to determine the work feed rates (cutting feedrates). This program word is used to establish feed rate values and precedes a numeric input for the feed amount in Inches per Minute (IPM) or, Inches per Revolution (IPR) and Meters per Minute (m/min) or Millimeters per Revolution (mm/rev) for metric programs. The value that is set by this command stays effective until changed by reentering a new value.

Feedrate Override

Feedrate Override allows control of the traverse feedrate by adjustment of a rotary dial. This function allows the control of the cutting feedrates defined by the F-word in the program by increasing or decreasing the percentage of the value entered in the program. It can also be used to control feedrates during jog mode function.

Fixed-Cycle

See Canned Cycle

G-Codes

Preparatory functions (G-Codes) are programmed with an address G, typically followed by two digits, to establish the mode of operation in which the tool moves.

Glossary

Incremental Dimension

An Incremental Dimension is a position within the coordinate system in which each numerical value is taken from the previous point.

Jog

Activating the JOG feed mode allows the selection of manual feeds along a single axis X, Y, or Z (rotary axes may be jogged, as well). With the mode activated, use the Axis/Direction buttons and the Speed/Multiply buttons to move the desired axis at the chosen feed rate (in/min or mm/min) and amount.

Linear Interpolation

This function allows programming of one, two or three axes simultaneously, that allows movement along a straight-line path, or at an angle in plane or space.

Machine Home

A reference position located within the machine tool working envelope, that is determined by the manufacturer, in order to establish a measurement system for the machine.

Machine Lock

An operation panel control, usually a button or a toggle switch, that allows the operator to lock all of the axes movements in order to check long programs for errors. If an error is encountered during this process, an alarm will be displayed on the control.

Manual Data Input (MDI)

The MDI mode enables the automatic control of the machine, using information entered in the form of blocks through the control panel without interfering with the basic program.

Miscellaneous Function

Miscellaneous functions are used to command various auxiliary operations such as activating coolant flow (M08) or starting clockwise spindle rotation (M03). The code consists of the letter M, typically followed by two digits. The M-Code is normally the last entry in a block.

Modal Commands

Modal commands remain in effect until they are replaced by another command from the same group. The F-Word is modal as are many G-Codes.

Glossary

Origin

A starting point for the coordinate system used to machine parts; a fixed point on a blueprint from which dimensions are taken.

Polar Coordinate System

A rotational coordinate system that locates points within a plane with respect to their distance from a fixed point of origin or pole by angular and radial values.

Preparatory Function (G-Word)   See G-Codes

Quadrant

A quadrant is one fourth (1/4) of a 2 dimensional grid in a plane of the rectangular coordinate system for measurement. It also represents an arc of 90° that is one fourth of a circle. Quadrant 1 is located in the upper right corner and 2 through 4 proceed in the counterclockwise direction.

Rectangular Coordinate System   See Cartesian Coordinates
Right Hand Rule

Using the right hand with palm up, the thumb will be pointing in the positive linear axis direction for X, then the little and ring fingers of the hand are folded over to touch the palm, the middle finger is allowed to point upwards (positive Z axis) and the index finger is pointing in the Y axis positive direction (vertical machine orientation).

Sequence Number (N-Word)

Sequence numbers (also called the block numbers) are identified by the letter N and are followed by one to five digits. A block number provides easier access to information contained in the program. The arrangement of block numbers in a given program can be random, but typically is sequenced in increments of one, two, five or ten. The most common step increments are five or ten. Block or sequence numbers can be omitted from a block (except in special cases, such as in some lathe cycles). The logical location then, for sequence numbers is at tool changes enabling the restart of that tool.

Single Block

The execution of a single block of information in the program is initiated by activating a switch or button on the control panel. While in this mode, each time the cycle start button is pressed, only one block of information will be executed.

Sub-Program

The subprogram is a subordinate program to the main program. It is registered in the controller memory with the letter O, followed by a four or five-digit number, same as the main program. In the main program, a subprogram is called by

Glossary

using the M98 function with the P-address to identify the sub-program number and then M99 (called for in the subprogram), is the function that ends a subprogram. Subprograms greatly simplify programming and decrease the amount of data that must be placed into the controller memory.

Tool Changer

A mechanical apparatus used to automatically change cutting tools by program control on CNC machines. The tool changer may be a magazine-type, with random access or, a carousel. The magazine-type uses a device similar to a robot arm that transfers tools from the magazine to the spindle and vice versa.

Tool Length Offset (TLO)

Tool Length Offsets "TLO", are called in the milling program by the H word. The measured values representing the difference between the spindle face gage line and the tool tip are input into corresponding offset registers and are needed for proper positioning of the tool along the Z axis.

U-Axis An additional linear axis parallel to X axis.

V-Axis An additional linear axis parallel to Y axis.

W-Axis An additional linear axis parallel to Z axis.

Work envelope

The working envelope is the maximum area of machining travel for each two axes in all four directions.

Workpiece Zero

A starting point, work piece zero is the point from which dimensions on the work piece are established. Sometimes referred to as part zero.

X-Axis

The axis of motion that is always horizontal and parallel to the machine tool table. For a vertical milling machine this axis moves left or right.

Y-Axis

The axis of motion that is perpendicular to both X and Z axes in relation to the machine tool table. For a vertical milling machine this axis moves forward and backward.

Z-Axis

The machine tool axis of motion that is always parallel to the primary spindle. For a vertical milling machine this axis moves up and down.

Index

A
A-Axis, 22, 30, 377
Absolute Coordinate System, 28, 264
Absolute Dimensioning System, 24
Address, 30-32, 377
Address Characters, 31
Address Searching, 68
Altering Program Words, 69
AOS, Algebraic Order System, 328, 375
APT, Automatically Programmed Tools, 342, 375
ATC, Automatic Tool Change, 7, 227, 347, 351, 375
ANSI, American National Standards Institute, 375
ASCII, American Standard Code for Information Interchange, 375
ASME, American Society of Mechanical Engineers, 24, 375

B
B-Axis, 22, 30, 377
Block, 98, 377
BCD, Binary Coded Decimal, 375

C
C-Axis, 22, 30, 378
Canned Cycles, 244-262
Cartesian Coordinates, 17, 378
Common Operation Procedures, 71-77
Control Panel Descriptions, 46-50
Coordinate Systems, 17-29
Cutter Diameter Compensation, 229-240
CAD, Computer Aided Design, 375
CAM, Computer Aided Manufacturing, 375
CAD/CAM, Computer Aided Design & Computer Aided Manufacturing, 193, 319-338, 341, 375
CIM, Computer-Integrated Manufacturing, 375
CW, Clockwise, 17, 350, 375
CCW, Counterclockwise, 17, 350, 375
CD-ROM, Compact Disc-Read Only Memory, 375
CD-RW, Compact Disc-Re-Writable, 375
CDC, Cutter Diameter Compensation, 7, 63, 64, 231-240, 375
CRC, Cutter Radius Compensation, 375
CNC, Computer Numerical Control, 375
CMM, Coordinate Measuring Machine, 375
CPU, Central Processing Unit, 352, 375
CRT, Cathode Ray Tube, 16, 46-48, 375

D
D, Tool Radius Offset Amount, 30, 63, 230, 336
D, Depth of Cut for Multiple Repetitive Cycles, 122, 124
Deleting a Program Word, 69
Dry Run, 16, 38-39, 65, 342, 378
DNC, Direct Numerical Control, 37, 66, 375
DVD, Digital Video Disc, 375
DXF, Drawing Exchange Format, 321, 375

E
E, Precise designation of thread lead, 30, 112, 116
Emergency Stop (E-stop), 36, 72
EDM, Electronic Discharge Machine, 375
EIA, Electronics Industries Association, 54, 71, 342, 347, 375
EOB, End of Block 31, 37, 49, 55, 97, 203, 375, 379

F
F, Feed rate, 30, 87, 88, 194
F, Precise Designation of Thread Lead, 30, 112, 116
F-Word, 103, 194, 213, 379
FMS, Flexible Manufacturing System, 375

G
General Information 3-11
G, Preparatory Functions, 30, 84, 196-198
G-Code, 29, 53, 102, 195, 211, 338-351, 375, 379
G00, Rapid Traverse, 84, 194, 196, 198, 230, 273
G01, Linear Interpolation, 84, 194, 196, 198, 230
G02, Circular Interpolation Clockwise, 84, 104, 214-222
G03, Circular Interpolation Counterclockwise 104, 214-222
G04, Dwell, 84, 196, 222, 262
G09, Exact Stop, 84, 196, 224
G15, Polar Coordinate Cancellation, 196, 224
G16, Polar Coordinate System, 196, 225
G17, X Y Plane Selection, 196, 214, 225
G18, X Z Plane Selection, 196, 215, 225
G19, Y Z Plane Selection, 196, 215, 224
G20, Input in Inches, 29, 84, 194, 196, 211, 227
G21, Input in Millimeters, 29, 84, 194, 196, 211, 227
G22, Stored Stroke Limit ON, 84, 196, 227
G23, Stored Stroke Limit OFF, 84, 196, 227
G25, Spindle Speed Fluctuation Detection ON, 84
G26, Spindle Speed Fluctuation Detection OFF, 84

Index

G27, Reference Point Return Check, 84, 109, 196, 227
G28, Reference Point Return, 84, 110, 149, 196, 227-229, 335
G29, Return from Reference Point, 84, 111, 196, 228
G30, Return to Second, Third and Fourth Reference Point, 84, 196, 229, 335
G32, Thread Cutting, 84, 111-113,
G37, Automatic Tool Length Measurement, 196
G40, Cutter Radius Compensation Cancellation, 84, 110, 148-150, 186, 196, 208, 211, 227, 229-240
G41, Cutter Radius Compensation Left, 84, 110, 148-150, 186, 196, 212, 220, 227, 229-240, 336
G42, Cutter Radius Compensation Right, 84, 110, 148-150, 186, 196, 212, 220, 227, 230, 336
G43, Positive Tool Length Offset Compensation, 196, 206
G44, Negative Tool Length Offset Compensation, 197, 208, 240
G49, Tool Length Offset Compensation Cancel, 197, 208, 211, 240
G50 Coordinate System Setting, 26, 28, 59, 84, 92, 94-96, 115
G50 Maximum Spindle Speed Setting, 84, 88, 90, 182
G52, Local Coordinate System, 84, 197, 225
G54 through G59, Work Coordinate Systems, 26, 28, 57, 84, 197, 202, 203, 241-243, 335, 347
G70, Finishing Cycle, 84, 127
G71, Stock Removal in Turning, 84, 120, 122
G72, Stock Removal in Facing, 84, 124
G73, Pattern Repeating, 84, 124-125
G73, High Speed Peck Drilling Cycle, 197, 246
G74, Peck Drilling Cycle, 84, 129
G74, Reverse Tapping, 197, 247
G75, Groove Cutting Cycle, 84, 130
G76, Fine Boring Cycle, 197, 247
G76, Multiple Thread Cutting Cycle, 84, 112, 131
G80, Canned Cycle Cancellation, 84, 197, 211, 248
G81, Drilling Cycle, Spot Drilling, 197, 209, 212, 248
G82, Counter Boring, "Chip Break" Cycle, 197,250
G83, Deep Hole Drilling Cycle, 71, 84, 197, 252
G84, Tapping Cycle, 84, 197, 254,
G85, Reaming Cycle, 197, 257
G86, Boring Cycle, 84, 197, 257
G87, Back Boring Cycle, 197, 261
G88, Boring Cycle, 197, 261
G89, Boring Cycle, 197, 262

G90, Absolute Programming Command, 197, 199, 208, 211, 263, 265
G90, Outer/Inner Diameter Turning Cycle, 84, 118
G91, Incremental Programming Command, 198, 199, 211, 212, 228, 229, 265
G92, Thread Cutting Cycle, 84, 113-116
G92, Work Coordinate System, 26, 28, 59, 201, 206, 241, 242, 244, 335
G94, Face Cutting Cycle, 84, 119
G96, Constant Surface Speed, 84, 88-90, 198
G97, Constant Surface Speed Cancellation, 84, 88-89, 113, 198
G98, Canned Cycle Initial Level Return, 198, 244-262
G98, Feed Per Minute, 84, 87, 88
G99, Feed Per Minute, 84, 87-88
G99, Canned Cycle Initial R-Level Return, 198, 244-262
G99, Feed Per Revolution, 84, 87
GB, Gigabit, 375
GD&T, Geometric Dimensioning & Tolerancing, 375
GHz, Gigahertz, 375
GUI, Graphical User Interface, 322, 341, 375

H

H, Tool Length Offset Number, 30, 63, 206, 212, 240, 241, 273
HP, Horsepower, 376
HSS, High Speed Steel, 376

I

I, Incremental X Coordinate for Arc Center, 30, 104-107, 215-220
I, Parameter of Fixed Cycle, 30, 117-119, 122, 124-126, 130, 133
Incremental Coordinate System, 28, 264, 335
ID, Inside Diameter, 376
IGES, Initial Graphics Exchange Specification, 321, 376
IPM, Inches Per Minute (also in/min), 10, 376
IPR, Inches Per Revolution (also in/rev), 10, 376
ISO, International Standards Organization, 54, 71, 342, 347, 376

J

J, Incremental Y Coordinate for Arc Center, 30, 215-219

K

K, Incremental Z Coordinate for Arc Center, 30, 104-107, 215-217, 220
K, Parameter of Fixed Cycle, 30, 119, 126, 130, 133, 245, 257
KW, Kilowatt, 376

Index

L

L, Number of Repetitions, 30, 99, 245
LAN, Local Area Network, 376
LED, Liquid Emitting Diode, 36, 40, 51, 52, 225, 267, 268, 376
LCD, Liquid Crystal Display, 47, 376

M

M, Miscellaneous Function (M-codes), 30, 80-83, 100, 198, 199, 376, 380
Machining Center Coordinate Systems, 199-89
Machining Center Example Programs, 263-313
Machining Center Preparatory Functions, 212-228
Machining Center Program Creation, 203-211
Machine Home, 24-25, 59, 380
Machine Zero, 24-26, 51, 92, 111, 202, 206, 208, 227, 242, 264, 348
Manual Data Input (MDI), 36, 37, 47, 49, 55, 56, 70, 267, 376, 380
Manual Pulse Generator (MPG), 40, 42, 201
Multiple Repetitive Cycles, 120-133
M00, Program Stop, 39, 40, 80, 81, 164, 198
M01, Optional Stop, 38, 73, 80, 81, 198, 209,
M02, Program End Without Rewind, 80, 82, 198, 206
M03, Spindle ON Clockwise, 37, 80, 88, 198, 248, 257, 261
M04, Spindle ON Counterclockwise, 80, 88, 198, 255
M05, Spindle OFF, 80, 198, 255, 257
M06, Tool Change, 80, 193, 198
M07, Mist Coolant ON, 80, 198
M08, Flood Coolant ON, 80, 198
M09, Coolant OFF, 80, 198, 209
M10, Chuck Close, 80, 82
M11, Chuck Open, 80, 82
M12, Tailstock Quill Advance, 80, 82
M13, Tailstock Quill Retract, 80, 82
M17, Rotation of Tool Turret Forward, 80, 83
M18, Rotation of Tool Turret Backward, 80, 83
M19, Spindle Orientation, 198, 247, 261
M21, Tail Stock Direction Forward, 80, 83
M21, Mirror Image X Axis, 198, 280
M22, Tail Stock Direction Backwards, 80, 83
M22, Mirror Image Y Axis, 198, 280
M23 Thread Finishing with Chamfer, 80, 83, 115
M23, Mirror Image Cancellation, 198, 280
M24, Thread Finishing with Right Angle, 80, 115
M30, Program End with Rewind, 56, 80, 83, 198, 206, 345
M98, Subprogram Call, 80, 98, 99, 198, 204
M99, Return to Main Program from Subprogram, 56, 80, 98, 198, 204
MB, Megabit, 46, 376
MCU, Machine Control Unit, 7, 376
MHz, Megahertz, 322, 376
Mm, Millimeters, 376

N

N, Sequence or Block Number, 30, 98, 203

O

O, Program Number, 30, 97-98, 203-204
Operation Panel Descriptions, 35-45
Operations Performed at the CNC Control, 51-71
Origin, 19-28, 48, 248, 323-325, 381
OD, Outside Diameter, 376

P

P, Subprogram Number Call, 30, 98-99, 204-205, 250, 252
P, Dwell Time Specification, 30, 108, 222
P, Start Sequence Number for Multiple Repetitive Cycles, 30, 127
Polar Coordinate System, 21, 381
Preparatory Functions (G-Codes), 83-84, 102-111, 381
Program Format, 29-30
Programming Coordinate Systems, 90-96
Programming CNC Machining Centers, 194-198
Programming CNC Turning Centers, 82-89
Program Creation, 97-101
Programming Examples For Lathes, 150-188
Process Planning for CNC, 11-16
PC, Personal Computer, 15, 66, 321-322, 352, 376
PCMCI, Portable Computer Memory Card Interface, 46, 54, 66, 376
PLC, Programmable Logic Controller, 376
PSI, Pounds Per Square Inch, 4, 376

Q

Q, Depth of Cut in Multiple Repetitive Cycles, 30, 129, 132
Q, Last Sequence Number for Multiple Repetitive Cycles, 30, 120, 125, 127
Q, Depth of Cut for Canned Cycles, 30, 245-246, 252-254
Quadrant, 21, 381

Index

R

R, Tool Nose Radius Offset Amount, 59
R, Radius Designation for Arc, 30, 105-107, 217, 219
R, Point R for Canned Cycles, as a Reference Return Value, 30, 245-262
Radial Offset Vector, 232-235
Rectangular Coordinate System, see Cartesian Coordinate System, 381
Right-Hand Rule, 18, 381
RAM, Random Access Memory, 322, 376
RPM, Revolutions Per Minute(also rev/min or r/min), 10, 44, 55, 89, 182, 326, 335, 343, 376
RISC, Reduced Instruction Set Computer, 352, 376
ROM, Read Only Memory, 376
R-8, Taper Designation 8, 376
RS232, Industry Standard Cabling Interface, 15, 16, 37, 66 376

S

S, Spindle-Speed Function, 30, 88-90, 195
Sequence Number (N-Word), 98, 381
Sequence Number Searching, 68
Simple Cutting Cycles, 112-119
SFM, Surface Feet per Minute, 376
SME, Society of Manufacturing Engineering, 376

T

T, Tool Function, 30, 85, 193-194
Tool Length Offset (TLO), 7, 32, 63-64, 376, 382
Tool Tip Orientation (T), 148, 150, 186
Tool Nose Radius, 134-149
TNRC, Tool Nose Radius Compensation, 148, 376
Three Dimensional Coordinate System, 19, 351
Two Dimensional Coordinate System, 19, 351

U

U, Additional Linear Axis Parallel to X axis, 30, 382
U, Dwell Time Specification, 108
U, Parameter for Multiple Repetitive Cycles, 120, 124-129

V

V axis, 30, 382

W

W axis, 30, 382
W, Parameter for Multiple Repetitive Cycles, 120, 121, 126
Work Coordinate Systems, 241-89
Word Searching, 68
Workpiece Zero, 24-27, 92, 200-202, 206, 208, 242, 382

X

X, X axis, 30, 382
X, Dwell Time Specification, 108

Y

Y, Y axis, 30, 382

Z

Z, Z, 30, 382

Answer Key to Study Questions

Part 1, **1.** T, **2.** c, **3.** b, **4.** c, **5.** d, **6.** a, **7.** F, **8.** 3, **9.** T, **10.** T, **11.** X0Y0, X0Y5.0, X2.5Y5.0, X4.0Y3.5, X4.0Y0, X0Y0 **12.** X0Y0, Y5.0, X2.5, X1.5Y-1.5, Y-3.5, X-4.0 **13.** Daily

Part 2, **1.** T, **2.** Position (All) or Program (Check), **3.** Active commands, **4.** Pressing the Input key enters a whole number where +Input enters an incremental amount, **5.** c, **6.** d,**7.** T, **8.** b, **9.** a, **10.** Manual Data Input, **11.** Offset, **12.** T, **13.** Jog

Part 3, **1.** c, **2.** T, **3.** The first two digits refer to the geometry offset and the second two are to the wear offset. **4.** b, **5.** d, **6.** T, **7.** b, **8.** T, **9.** a, **10.** b, **11.** c, **12.** b, **13.** b, **14.** a negative sign, **15.** T, **16.** d, **17.** b, **18.** d, **19.** b

Part 4, **1.** b, **2.** d, **3.** c, **4.** d, **5.** b, **6.** d, **7.** c, **8.** T, **9.** d, **10.** c, **11.** d, **12.** a, **13.** T, **14.** F, **15.** b, **16.** b, **17.** a, **18.** c, **19.** d, **20.** a, **21.** a. M00, b. M01, c. M30, d. M03, e. M04, f. M05, g. M08, h. M09, i. M19, j. M98, k. M99, l. M06, **22.** a. G01, b. G02, c. G00, d. G04, e. G90, f. G91, g. G80, h. G83, i. G81, j. G41, k. G42, l. G40, m. G28, n. G20, o. G21, p. G84, q. G92, r. G54, s.G43, **23.** a. O, b. N, c. G, d. M, e. X, f. Y, g. Z, h. R, i. F, j. S, k. L, l. T, m. H, n. D o. A, p. B, q. C, r. P, s. Q, t. I, u. J, v. K **24.** c, **25.** T, **26.** T